FreeSWITCH 1.6 Cookbook

Over 45 practical recipes to empower you with the latest FreeSWITCH 1.6 features

Anthony Minessale II

Michael S Collins

Giovanni Maruzzelli

[PACKT] PUBLISHING

open source
community experience distilled

BIRMINGHAM - MUMBAI

FreeSWITCH 1.6 Cookbook

First published: February 2012

Second Edition: July 2015

Production reference: 1170715

Published by Packt Publishing Ltd.
Livery Place
35 Livery Street
Birmingham B3 2PB, UK.

ISBN 978-1-78528-091-7

www.packtpub.com

Credits

About the Authors

Anthony Minessale II is the primary author and founding member of the FreeSWITCH open source softswitch. He has spent almost 20 years working with open source software. In 2001, he spent a great deal of time as an Asterisk developer and authored numerous features and fixes to that project. Anthony started coding a new idea for an open source voice application in 2005. The FreeSWITCH project was officially opened to the public on January 1, 2006. In the years that followed, Anthony has actively maintained and led software development for this project.

Michael S Collins is a telephony and open source software enthusiast. Having worked as a PBX technician for 5 years and the head of IT for a call center for more than 9 years, he is a PBX veteran. He is an active member of the FreeSWITCH community and has coauthored *FreeSWITCH Cookbook*, by Packt Publishing in 2012. Michael lives in Central California with his wife and two children.

Giovanni Maruzzelli (available at `OpenTelecom.IT`) is heavily engaged with FreeSWITCH. In it, he wrote interfacing with Skype and cellular phones. He's a consultant in the telecommunication sector, developing software and conducting training courses for FreeSWITCH, SIP, WebRTC, Kamailio, and OpenSIPS.

An Internet technology pioneer, he was the cofounder of Italia Online in 1996. It is the most popular Italian portal and consumer ISP. Also, he was the architect of its Internet technologies (`www.italiaonline.it`). Then, Giovanni was the supervisor of Internet operations and the architect of the first engine for paid access to `ilsole24ore.com`, the most read financial newspaper in Italy, and its databases (migrated from the mainframe).

After that, he was the CEO of the venture-capital-funded company Matrice, developing telemail unified messaging and multiple-language phone access to e-mail (text to speech). He was also the CTO of the incubator-funded company Open4, an open source managed applications provider.

For 2 years, Giovanni worked in Serbia as an Internet and telecommunication investment expert for IFC, an arm of The World Bank.

Since 2005, he has been based in Italy and serves ICT and telecommunication companies worldwide.

For my first book, first and foremost I want to thank my late guru, Marco Amante, who taught me to work, to write, "computerism," the most important things in life, as well as their order. Thanks, master!

Also, I want to thank Gisella Genna for her continued, incredible, and precious support and love. Thanks, my beloved!

About the Reviewers

Ayobami Adewole is a software engineer who has spent the last 4 years working on various cutting-edge VoIP solutions using FreeSwitch IP-PBX. He graduated with a degree in computer science from the Ladoke Akintola University of Technology, Ogbomoso, Nigeria. He has developed applications and authored solutions for different projects, such as land administration and geographical information systems, enterprise-level application integrations, and unified communication and software applications for the education and business sectors. In his spare time, he enjoys experimenting with new technologies.

> I would like to acknowledge my parents for instilling in me the culture of discipline and hard work. Also, my gratitude goes to my partner for always encouraging me. Lastly, I want to thank everyone who had assisted me and given me the opportunity to pursue my passion for technology.

Josh Richesin is a cofounder of Sureline Broadband, a broadband and telephony company based in Madras, Oregon, USA. He also has his own private consulting practice, primarily for service providers. He has been in the telecommunication industry for his entire career and has always had a focus on large, mission-critical networks in the service providers' realm. You can contact Josh and read his blog at www.joshrichesin.com.

Gabe Shepard is a master developer at Star2Star Communications and has been dabbling in the VoIP world for nearly a decade. Prior to that, he focused on Linux systems administration and web development. He currently lives in Sarasota, Florida, USA, with his wife, Kelli, and their two daughters, Charlotte and Lucille. You can contact him at `extrema.net`.

Brian Wiese has always been passionate about technology and enjoys sharing his knowledge with others. Employed by an educational software and consulting company serving the requirements of schools throughout the United States, he continues to fulfill his personal commitment in finding sustainable and cost-effective solutions for educators. Brian's more than 10 years of experience in telephony helps him identify winning strategies for VoIP integration.

First, I wish to thank the authors of this cookbook for giving me the opportunity to review it after reviewing *FreeSWITCH 1.2*. Again, I'd like to recognize the accomplishments of Anthony and the rest of the community in making FreeSWITCH an amazing telephony platform. Finally, I send a most sincere thank you to my parents and friends for all of their encouragement, support, and inspiration.

www.PacktPub.com

Support files, eBooks, discount offers, and more

For support files and downloads related to your book, please visit www.PacktPub.com.

Did you know that Packt offers eBook versions of every book published, with PDF and ePub files available? You can upgrade to the eBook version at www.PacktPub.com and as a print book customer, you are entitled to a discount on the eBook copy. Get in touch with us at service@packtpub.com for more details.

At www.PacktPub.com, you can also read a collection of free technical articles, sign up for a range of free newsletters and receive exclusive discounts and offers on Packt books and eBooks.

https://www2.packtpub.com/books/subscription/packtlib

Do you need instant solutions to your IT questions? PacktLib is Packt's online digital book library. Here, you can search, access, and read Packt's entire library of books.

Why Subscribe?

- ▸ Fully searchable across every book published by Packt
- ▸ Copy and paste, print, and bookmark content
- ▸ On demand and accessible via a web browser

Free Access for Packt account holders

If you have an account with Packt at www.PacktPub.com, you can use this to access PacktLib today and view 9 entirely free books. Simply use your login credentials for immediate access.

Table of Contents

Preface

FreeSWITCH is increasingly becoming the "serious choice" for companies to base their products and offerings on. Its usage is widespread, scaling from Raspberry Pis to "Big Irons" in the data center.

There is a growing need for books and training, and with Packt Publishing, we decided to begin serving this burgeoning demand. This cookbook is a primer; then there will be a *Mastering FreeSWITCH* book, followed by a new edition of the *classic* FreeSWITCH book.

Obviously, nothing can beat a training camp or codeveloping in collaboration with an old hand, but these FreeSWITCH titles will form the basis on which a company or a consultant can begin embracing, deploying, and implementing FreeSWITCH.

This book is a complete update, rewrite, and integration of the old FreeSWITCH cookbook. This new edition covers FreeSWITCH 1.6.x, and a lot of new ground.

All the examples here have been updated and tested with the new FreeSWITCH series, while a new section has been added about connecting to Skype, and two entire chapters are on WebRTC and Lua programming.

Anthony Minessale II, Giovanni Maruzzelli

July 5 2015

What this book covers

Chapter 1, Routing Calls, shows that getting calls from one endpoint to another is the primary function of FreeSWITCH. This chapter discusses techniques of efficiently routing calls between phones and service providers.

Chapter 2, Connecting Telephones and Service Providers, assists in quickly getting your FreeSWITCH server connected to other VoIP devices. Telephones and service providers have specific requirements for connecting to FreeSWITCH.

Chapter 3, *Processing Call Detail Records*, discusses a number of ways to extract CDR data from your FreeSWITCH server. Call detail records, or CDRs, are very important for businesses.

Chapter 4, *External Control*, presents a number of real-world examples of controlling FreeSWITCH from an external process. FreeSWITCH can be controlled externally by the powerful and versatile event socket interface.

Chapter 5, *PBX Functionality*, is the largest chapter in this book. This chapter shows how to deploy features such as voicemail, conference calls, faxing, IVRs, and more, which most telephone systems have, in a FreeSWITCH server.

Chapter 6, *WebRTC and Mod_Verto*, features the new disruptive technology that allows real-time audio/video/data-secure communication from hundreds of millions of browsers. FreeSWITCH is ready to serve as a gateway and an application server.

Chapter 7, *Dialplan Scripting with Lua*, covers Lua, the scripting language of choice for programming complex logic in FreeSWITCH. Accessing databases, calling web servers, and interacting with user's choices now becomes easy.

What you need for this book

FreeSWITCH 1.6 Cookbook is an essential addition to any VoIP administrator's or WebRTC developer's library. PBX implementers will also gain from the thoroughly distilled recipes presented here.

Whether you are a FreeSWITCH expert or are just getting started, this book will take your skills to the next level.

Who this book is for

FreeSWITCH 1.6 Cookbook is written for anyone who wants to learn more about using FreeSWITCH in production. The information is presented in such a way that you can get up and running quickly. The cookbook approach eschews much of the foundational concepts, and instead focuses on discrete examples that illustrate specific features. If you need to implement a particular feature as quickly as possible, then this book is for you.

Sections

In this book, you will find several headings that appear frequently (Getting ready, How to do it, How it works, There's more, and See also).

To give clear instructions on how to complete a recipe, we use the following sections.

Getting ready

This section tells you what to expect in the recipe, and describes how to set up any software or any preliminary settings required for the recipe.

How to do it...

This section contains the steps required to follow the recipe.

How it works...

This section usually consists of a detailed explanation of what happened in the previous section.

There's more...

This section consists of additional information about the recipe in order to make you more knowledgeable about the recipe.

See also

This section provides helpful links to other useful information for the recipe.

Conventions

In this book, you will find a number of text styles that distinguish between different kinds of information. Here are some examples of these styles and an explanation of their meaning.

Code words in text, database table names, folder names, filenames, file extensions, pathnames, dummy URLs, user input, and Twitter handles are shown as follows: "Many of the techniques employed in `Local_Extension` are discussed in this chapter."

A block of code is set as follows:

```
<include>
  <extension name="public_did">
    <condition field="destination_number"
    expression="^(8005551212)$">
      <action application="set" data="domain_name=$${domain}"/>
      <action application="transfer" data="1000 XML default"/>
    </condition>
  </extension>
</include>
```

When we wish to draw your attention to a particular part of a code block, the relevant lines or items are set in bold:

```
<include>
  <extension name="public_did">
    <condition field="destination_number"
    expression="^(8005551212)$">
      <action application="set" data="domain_name=$${domain}"/>
      <action application="transfer" data="1000 XML default"/>
    </condition>
  </extension>
</include>
```

Any command-line input or output is written as follows:

```
perl -MCPAN -e 'install Regexp::Assemble'
```

New terms and **important words** are shown in bold. Words that you see on the screen, for example, in menus or dialog boxes, appear in the text like this: "You should see an application named **directory** in the list."

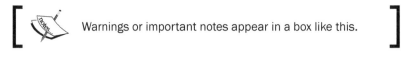

Warnings or important notes appear in a box like this.

Tips and tricks appear like this.

Reader feedback

Feedback from our readers is always welcome. Let us know what you think about this book— what you liked or disliked. Reader feedback is important for us as it helps us develop titles that you will really get the most out of.

To send us general feedback, simply e-mail feedback@packtpub.com, and mention the book's title in the subject of your message.

If there is a topic that you have expertise in and you are interested in either writing or contributing to a book, see our author guide at www.packtpub.com/authors.

Customer support

Now that you are the proud owner of a Packt book, we have a number of things to help you to get the most from your purchase.

Downloading the example code

You can download the example code files from your account at `http://www.packtpub.com` for all the Packt Publishing books you have purchased. If you purchased this book elsewhere, you can visit `http://www.packtpub.com/support` and register to have the files e-mailed directly to you.

Downloading the color images of this book

We also provide you with a PDF file that has color images of the screenshots/diagrams used in this book. The color images will help you better understand the changes in the output. You can download this file from `https://www.packtpub.com/sites/default/files/downloads/B04231_ColoredImages.pdf`.

Errata

Although we have taken every care to ensure the accuracy of our content, mistakes do happen. If you find a mistake in one of our books—maybe a mistake in the text or the code—we would be grateful if you could report this to us. By doing so, you can save other readers from frustration and help us improve subsequent versions of this book. If you find any errata, please report them by visiting `http://www.packtpub.com/submit-errata`, selecting your book, clicking on the **Errata Submission Form** link, and entering the details of your errata. Once your errata are verified, your submission will be accepted and the errata will be uploaded to our website or added to any list of existing errata under the Errata section of that title.

To view the previously submitted errata, go to `https://www.packtpub.com/books/content/support` and enter the name of the book in the search field. The required information will appear under the **Errata** section.

Piracy

Piracy of copyrighted material on the Internet is an ongoing problem across all media. At Packt, we take the protection of our copyright and licenses very seriously. If you come across any illegal copies of our works in any form on the Internet, please provide us with the location address or website name immediately so that we can pursue a remedy.

Please contact us at copyright@packtpub.com with a link to the suspected pirated material.

We appreciate your help in protecting our authors and our ability to bring you valuable content.

Questions

If you have a problem with any aspect of this book, you can contact us at questions@packtpub.com, and we will do our best to address the problem.

1
Routing Calls

In this chapter, we will discuss routing calls in various scenarios, as follows:

- ▸ Internal calls
- ▸ Incoming DID (also known as DDI) calls
- ▸ Outgoing calls
- ▸ Ringing multiple endpoints simultaneously
- ▸ Ringing multiple endpoints sequentially (simple failover)
- ▸ Advanced multiple endpoint calling with enterprise originate
- ▸ Time-of-day routing
- ▸ Manipulating SIP To: headers on registered endpoints to reflect DID numbers

Introduction

Routing calls is at the core of any FreeSWITCH server. There are many techniques for accomplishing the surprisingly complex task of connecting one phone to another. However, it is important to make sure that you have the basic tools necessary to complete this task.

The most basic component of routing calls is the **dialplan**, which is essentially a list of actions to perform depending upon which digits were dialed (as we will see in some of the recipes in this book, there are other factors that can affect routing of calls). The dialplan is broken down into one or more **contexts**. Each context is a group of one or more **extensions**. Finally, each extension contains specific **actions** to be performed on the call. The dialplan processor uses **regular expressions**, which are a pattern-matching system used to determine which extensions and actions to execute.

To make best use of the recipes in this chapter, it is especially important to understand how to use regular expressions and the three contexts in the default configuration.

Regular expressions

FreeSWITCH uses **Perl-compatible regular expressions** (**PCRE**) for pattern matching. Consider this dialplan excerpt:

```
<extension name="example">
<condition field="destination_number" expression="^(10\d\d)$">
<action application="log" data="INFO dialed number is [$1]"/>
```

This example demonstrates the most common uses of regular expressions in the dialplan: matching against the destination_number field (that is, the digits that the user dialed) and capturing, using parentheses, the matched value in a special variable named $1. Let's say that a user dials 1025. Our example extension will match 1025 against the ^(10\d\d)$ pattern and determine that this is indeed a match. All actions inside the condition tag will be executed. The action tag in our example will execute the log application. The log application will then print a message to the console, using the INFO log level, which will be in green text by default. The value in $1 is **expanded** (or **interpolated**) when printed:

```
2015-02-22 15:15:50.664585 [INFO] mod_dptools.c:1628 dialed number is
[1025]
```

Understanding these basic principles will help you create effective dialplan extensions.

 For more tips on using regular expressions, be sure to visit http://freeswitch.org/confluence/display/ FREESWITCH/Regular+Expression.

Important dialplan contexts in the default configuration

Contexts are logical groups of extensions. The default FreeSWITCH configuration contains three contexts:

- default
- public
- features

Each of these contexts serves a purpose, and knowing about them will help you leverage their value for your needs.

The default context

The most used context in the default configuration is the `default` context. All users whose calls are **authenticated** by FreeSWITCH will have their calls passing through this context, unless there have been modifications. Some common modifications include using ACLs or disabling authentication altogether (see the *The public context* section that follows). The `default` context can be thought of as **internal** in nature; that is, it services users who are connected directly to the FreeSWITCH server, as opposed to outside callers (again, see the *The public context* section).

Many characteristics related to PBX (Private Branch Exchange) are defined in the `default` context, as are various utility extensions. It is good to open `conf/dialplan/default.xml` and study the extensions there. Start with simple extensions such as `show_info`, which performs a simple `info` dump to the console, and `vmain`, which allows a user to log in to their voicemail box.

A particularly useful extension to review is `Local_Extension`. This extension does many things, as follows:

- ▸ Routes calls between internal users
- ▸ Sends calls to the destination user's voicemail on a no-answer condition
- ▸ Enables several in-call features with `bind_meta_app`
- ▸ Updates (via "hash" commands) the local calls database to allow call return and call pickup

Many of the techniques employed in `Local_Extension` are discussed in this chapter (see the *The features context* section for a discussion on the in-call features found in this extension).

The public context

The `public` context is used to route incoming calls that originate from outside the local network. Calls that initially come to the `public` context are treated as untrusted. If they are not specifically routed to an extension in the `default` context, then they are simply disconnected. As mentioned before, disabling authentication or using ACLs to let calls into the system will route them into the `public` context (this is a security precaution, which can be overridden if absolutely required). We will use the `public` context in the *Incoming DID (also known as DDI) calls* recipe.

The features context

The `features` context is used to make certain features available for calls that are in progress. Consider this excerpt from `Local_Extension` in `conf/dialplan/default.xml`:

```
<action application="bind_meta_app" data="1 b s
execute_extension::dx XML features"/>
```

This is just one of several features that are enabled for the recipient of the call. The `bind_meta_app` application listens on the audio stream for a touch-tone * followed by a single digit. The preceding example is a blind transfer. If the called user dials *1*, then the `execute_extension::dx XML features` command is executed. In plain words, this command says, "Go to the `features` context of the XML dialplan and execute the extension whose destination number is `dx`." In `conf/dialplan/features.xml`, there is the following extension:

```
<extension name="dx">
<condition field="destination_number" expression="^dx$">
  ...
```

The `dx` extension accepts some digits from the user and then transfers the caller to the destination that the user keyed in.

This process demonstrates several key points:

- Calls can be transferred from one dialplan context to another
- The `features` context logically isolates several extensions that supply in-call features
- The `bind_meta_app` dialplan application is one of the means of allowing in-call features

Understanding that *calls can flow from one context to another* even after they are in progress is an important concept to grasp when addressing your call routing scenarios.

Internal calls

Calling local extensions is very simple once you know what needs to happen. In this case, we will see how to add a new user and make their phone available for calling.

Getting ready

If you are using the default configuration, then users 1000 through 1019 are preconfigured, *both in the directory and the dialplan*. To add a user beyond this range to the directory, it is generally easier to run the `add_user` script, found in the FreeSWITCH source directory under `scripts/perl`. For example, to add user 1020 to the directory, launch this script from the FreeSWITCH source directory, specifying the user ID on the command line:

```
scripts/perl/add_user 1020
```

You can also specify a range of users:

```
scripts/perl/add_user --users=1020-1029
```

You will see a note about the number of users added to the directory. If you have the `Regexp::Assembly` CPAN module installed, then the script will also generate a couple of *sample regular expression patterns*, which you can then use in the dialplan. For our example, we added a range of users from 1020 to 1029 to the directory, and then we'll add them to the dialplan.

How to do it...

1. Open the `conf/dialplan/default.xml` file in a text editor. Locate the `Local_Extension` entry:

   ```
   <extension name="Local_Extension">
   <condition field="destination_number"
   expression="^(10[01][0-9])$">
   ```

2. Edit the expression in the `<condition>` tag to account for our new users. The `^(10[012][0-9])$` expression pattern will do what we need (look closely to see the difference). The new line will be as follows:

   ```
   <condition field="destination_number"
   expression="^(10[012][0-9])$">
   ```

3. Save the file and then execute `reloadxml` from `fs_cli`.

Downloading the example code

You can download the example code files for all Packt Publishing books you have purchased from your account at `http://www.packtpub.com`. If you purchased this book elsewhere, you can visit `http://www.packtpub.com/support` and register to have the files e-mailed directly to you.

How it works...

`Local_Extension` is the default dialplan entry that allows directory users to be called. Remember that *simply adding a user to the directory does not mean that the user can be dialed*. (However, it does usually mean that the user can make outbound calls.) So, in order for your new user to be reachable, you need to add their user ID to the dialplan. By default, `Local_Extension` has a regular expression that will match 1000, 1001, and so on up to 1019. After adding that number range, it is necessary to modify the regular expression to account for those new numbers. In our example, we added user IDs 1020 through 1029, so we need to match these. We use this regular expression:

```
^(10[012][0-9])$
```

This matches 1000 through 1029. Let's say we have added another block of user IDs with the range from 1030 to 1039. We can modify our regular expression to catch them as well:

```
^(10[0123][0-9])$
```

It is considered best practice not to add a large range of dialable numbers to `Local_Extension` without having the corresponding users in the directory. Doing so can make troubleshooting dialplan issues more difficult.

As a reminder, be sure to execute the `reloadxml` command each time you modify the regular expression (the changes you make to your XML configuration files will not take effect until they are loaded into the memory, which is what the `reloadxml` command does).

See also

▶   The *Creating users* section in *Chapter 5, PBX Functionality*

Incoming DID (also known as DDI) calls

Phone calls coming in from the **Public Switched Telephone Network** (**PSTN**) are often called DID or DDI calls. DID stands for **Direct Inward Dialing**, while DDI means **Direct Dial In**; both acronyms refer to the same thing. DID numbers are delivered by your telephone service provider. They can be delivered over VoIP connections (such as a SIP trunk) or via traditional telephone circuits, such as PRI lines. These phone numbers are sometimes called *DID numbers* or *external phone numbers*.

Getting ready

Routing a call requires two pieces of information: the phone number being routed and a destination for that phone number. In our example, we will use a DID number `8005551212`. Our destination will be user `1000`. Replace these sample numbers with the appropriate values for your setup.

How to do it...

1.   Create a new file in `conf/dialplan/public/` named `01_DID.xml`. Add this text to it:

```
<include>
<extension name="public_did">
<condition field="destination_number"
    expression="^(8005551212)$">
<action application="set" data="domain_name=$${domain}"/>
<action application="transfer" data="1000 XML default"/>
```

```
          </condition>
          </extension>
          </include>
```

2. Save the file and then execute `reloadxml` from `fs_cli`.

How it works...

All calls that come in to the FreeSWITCH server from outside (as well as internal calls that are not authenticated) are initially handled in the `public` context (dialplan contexts were discussed in more detail in this chapter's introduction) of the XML dialplan. Once the call hits the `public` context, we try to match the `destination_number` field. The `destination_number` is generally the DID number (see the *There's more...* section for some caveats). Once we match the incoming number, we set the `domain_name` channel variable to the default domain value, and then transfer the call to user 1000. (FreeSWITCH is domain-based in a way similar to e-mails. Most systems have only a single domain, though FreeSWITCH supports multiple domains.) The actual transfer happens with this dialplan entry:

```
<action application="transfer" data="1000 XML default"/>
```

In plain words, this tells FreeSWITCH to transfer the call to extension 1000 in the default context of the XML dialplan. The `default` context contains the `Local_Extension` that matches "1000" as `destination_number` and handles the calls to users' telephones.

There's more...

Keep in mind that the expression for `destination_number` must match what the provider sends to FreeSWITCH, not necessarily what the calling party actually dialed. There are providers that send DID information in various formats, such as these:

- 8005551212
- 18005551212
- +18005551212

The expression must match what the provider sends. One way to accomplish this is to have a few *optional characters in the pattern*. This pattern matches all three formats you just saw:

```
<condition field="destination_number"
expression="^\+?1?(8005551212)$">
```

The `\+?` value means "optionally match the literal + character," and the `1?` value means "optionally match the literal digit 1." Now our pattern will match all of the three formats that are commonly used in North America. (Technically, our pattern will also match `+8005551212`, but we are not concerned about that. However, the pedantic admin might be, so they can use the `^(\+1)?1?(8005551212)$` pattern instead.)

See also

▸ The *Configuring a SIP gateway* section in *Chapter 2, Connecting Telephones and Service Providers*

Outgoing calls

In order to make your system useful, you need a way to dial out to the "real world". This recipe will cover dialing out to the PSTN and allow you to connect to landlines, cellular phones, and so on. In this recipe, we'll make an extension that will allow an outbound call to any valid US number. We'll attempt to complete the call using the gateway named `our_sip_provider` (see the *Configuring an SIP Gateway* section in *Chapter 2, Connecting Telephones and Service Providers*).

Getting ready

Making outbound calls requires you to know the numbering format that your provider requires. For example, do they require all 11 digits for US dialing? Or will they accept 10? In our example, we're going to assume that our provider will accept a 10-digit format for US dialing (for example, without the international prefix "1").

How to do it...

Routing outbound calls is simply a matter of creating a dialplan entry:

1. Create a new file in `conf/dialplan/default/` named `outbound_calls.xml`. then add the following text:

```xml
<include>
<extension name="outbound_calls">
<condition field="destination_number"
    expression="^1?([2-9]\d{2}[2-9]\d{6})$">
<action application="bridge" data="sofia/gateway/our_sip_
provider/$1"/>
</condition>
</extension>
</include>
```

2. Save your XML file and issue the `reloadxml` command at `fs_cli`.

How it works...

Assuming you have a phone set up on the `default` context, your regular expression will match any `destination_number` that follows the US dialing format (10 or 11 digits), and send the call to `our_sip_provider` in a 10-digit format. The format in regexp is as follows: optional "1", then one digit between 2 and 9, then two digits, then one digit between 2 and 9, and finally six digits. Only the part after the optional digit "1" is captured by the parentheses and passed down in the `$1` variable.

There's more...

The regular expression matching in FreeSWITCH allows the privilege of having very powerful conditions. You can also match `caller_id_number` to route calls from a user calling from extension 1011 out to the second gateway called `our_second_sip_provider`, while everyone else will be sent through `our_sip_provider`. Consider the following alternative `outbound_calls.xml` file:

```
<include>
<extension name="outbound_calls_from_1011">
<condition field="caller_id_number" expression="^1011$"/>
<condition field="destination_number"
    expression="^1?([2-9]\d{2}[2-9]\d{6})$">
<action application="bridge"
      data="sofia/gateway/our_second_sip_provider/$1"/>
</condition>
</extension>
<extension name="outbound_calls">
<condition field="destination_number"
    expression="^1?([2-9]\d{2}[2-9]\d{6})$">
<action application="bridge"
      data="sofia/gateway/our_sip_provider/$1"/>
</condition>
</extension>
</include>
```

Note that we have two extensions. The first one tries to match the `caller_id_number` field to the value 1011. If it matches 1011, then the call gets sent to the `our_second_sip_provider` gateway. Otherwise, the next extension is matched and the call goes to the `our_sip_provider` gateway. Note that we use `$1` to capture the matching value in the conditions' expressions. In each case, we capture exactly 10 digits, which correspond to the area code (three digits), exchange (three digits), and phone number (four digits). These are **North American Numbering Plan (NANP)** numbers. The regular expressions used to capture the format of dialed digits vary, depending upon the country.

Regular expressions can be a challenge. There are a number of examples with explanations on the FreeSWITCH wiki. See `http://freeswitch.org/confluence/display/FREESWITCH/Regular+Expression` for further details.

See also

▶ The *Configuring an SIP phone to register with FreeSWITCH* and *Configuring an SIP gateway* sections in *Chapter 2, Connecting Telephones and Service Providers*

Ringing multiple endpoints simultaneously

FreeSWITCH makes it easy to ring multiple endpoints simultaneously within a single command.

Getting ready

Open `conf/dialplan/default.xml` in a text editor or create a new XML file in the `conf/dialplan/default/` subdirectory.

How to do it...

Add a comma-separated list of endpoints to your `bridge` (or `originate`) application. For example, to ring `userA@local.pbx.com` and `userB@local.pbx.com` simultaneously, use an extension like this:

```
<extension name="ring_simultaneously">
<condition field="destination_number" expression="^(2000)$">
<action application="bridge"
    data="{ignore_early_media=true}sofia/internal/
userA@local.pbx.com,sofia/internal/userB@local.pbx.com"/>
</condition>
</extension>
```

How it works...

Putting comma-separated endpoints into the argument to `bridge` causes all the endpoints in that list to be dialed simultaneously. It sounds simple; however, there are several factors to consider when ringing multiple devices simultaneously in a real environment. The `bridge` application will *connect the call to whoever sends the media first*. This *includes early media (ringing)*. To put this in other words, if you bridge a call to two parties and one party starts sending a ringing signal back to you, that will be considered media and the call will be connected to that party. *Ringing of the other phones will cease.*

If you notice that calls always go to a specific number on your list of endpoints versus ringing all numbers, or that all phones ring for a moment before ringing only a single number, it means that your call may be getting bridged prematurely because of early media. Notice that we added `ignore_early_media=true` at the beginning of the dial string. As its name implies, `ignore_early_media` tells the `bridge` application not to connect the calling party to the called party when receiving early media (such as a ringing or busy signal). Instead, `bridge` will only connect the calling party to the called party who actually answers the call. In most cases, it is useful to *ignore early media when ringing multiple endpoints* simultaneously.

There's more...

In some scenarios, you may also wish to ring specific devices for a limited amount of time. You can apply the `leg_timeout` parameter to each leg of the bridge to specify how long to ring each endpoint like this:

```
<action application="bridge"
data="[leg_timeout=20]sofia/internal/userA@local.pbx.com,
[leg_timeout=30]sofia/internal/userB@local.pbx.com"/>
```

In this example, the `userA` user's phone will ring for a maximum of 20 seconds, while the `userB` user's phone will ring for a maximum of 30 seconds.

Call legs and the leg_timeout variable

The `leg_timeout` variable is unique in the sense that it implies the ignoring of early media. When using the `leg_timeout` variable on each call leg in a `bridge` attempt, there is no need to explicitly use `{ignore_early_media=true}` in the `bridge` argument. For a more thorough discussion of using `{` and `}` (curly brackets) versus `[` and `]` (square brackets), see `http://freeswitch.org/confluence/display/FREESWITCH/Channel+Variables#ChannelVariables-ChannelVariablesinDialstrings`.

This method of calling multiple parties works well for a small number of endpoints. However, it does not scale to dozens or more users. Consider using a FIFO queue in such an environment (FreeSWITCH's `mod_fifo` is discussed at length at `http://freeswitch.org/confluence/display/FREESWITCH/mod_fifo`).

See also

▶ The *Ringing multiple endpoints sequentially (simple failover)* recipe that follows for an example of ringing a group of endpoints one at a time, which includes an expanded discussion of using call timeouts

Ringing multiple endpoints sequentially (simple failover)

Sometimes it is necessary to ring additional endpoints, but only if the first endpoint fails to connect. The FreeSWITCH XML dialplan makes this very simple.

Getting ready

Open `conf/dialplan/default.xml` in a text editor or create a new XML file in the `conf/dialplan/default/` subdirectory.

How to do it...

Add a pipe-separated list of endpoints to your `bridge` (or `originate`) application. For example, to ring `userA@local.pbx.com` and `userB@local.pbx.com` sequentially, use an extension like this:

```
<extension name="ring_sequentially">
<condition field="destination_number" expression="^(2001)$">
<action application="bridge"
    data="{ignore_early_media=true}sofia/internal/
userA@local.pbx.com|sofia/internal/userB@local.pbx.com"/>
</condition>
</extension>
```

How it works...

Putting pipe-separated endpoints in the argument to `bridge` (or `originate`) causes all the endpoints in that list to be dialed sequentially. The first endpoint on the list that is successfully connected will be bridged, and the remaining endpoints will not be dialed. There are several factors to consider when ringing multiple devices sequentially.

Notice that we added `ignore_early_media=true` at the beginning of the dial string. As its name implies, `ignore_early_media` tells the `bridge` application not to connect the calling party to the called party when receiving early media (such as a ringing or busy signal). Instead, `bridge` will only connect the calling party if the called party actually answers the call. In most cases, you will need to ignore early media when dialing multiple endpoints.

There's more...

Handling different failure conditions can be a challenge. FreeSWITCH has a number of options that let you tailor `bridge` and `originate` to your specific requirements.

Handling busy and other failure conditions

For example, when calling a user who is on the phone, one service provider might return SIP message 486 (USER_BUSY), whereas many providers might simply send a SIP 183 with SDP and a media stream with a busy signal. In the latter, how will the `bridge` application know that there is a failure if it is ignoring the early media that contains the busy signal? FreeSWITCH gives us a tool that allows us to *monitor* early media even while "ignoring" it.

Consider two very common examples of *failed* calls where the failure condition is signaled in-band:

- ▸ Calling a line that is in use
- ▸ Calling a disconnected phone number

These conditions are commonly communicated to the caller via specific sounds: busy signals and special information tones, or **SIT** tones. In order for the early media to be meaningful, we need to be able to listen for specific tones or frequencies. Additionally, we need to be able to specify that certain frequencies mean different kinds of failure conditions (this becomes important for reporting, as in call detail records or CDRs). The tool that FreeSWITCH provides us with is a special channel variable called `monitor_early_media_fail`. Its use is best illustrated with an example:

```
<action application="bridge" data="{ignore_early_media=true,
monitor_early_media_fail=user_busy:2:480+620!
destination_out_of_order:2:1776.7}sofia/internal/
userA@local.pbx.com|sofia/internal/userB@local.pbx.com"/>
```

Here, we have a `bridge` application that ignores early media and sets two failure conditions: one for `busy` and one for `destination_out_of_order`. We specify the name of the condition we are checking, the number of `hits`, and the frequencies to detect. The format for `monitor_early_media_fail` is as follows:

```
condition_name:number_of_hits:tone_detect_frequencies
```

The `user_busy` condition is defined as `user_busy:2:480+620`. This condition looks for both 480 Hz and 620 Hz frequencies (which is the USA busy signal), and if they are detected twice, then the call will *fail*. The exclamation mark (`!`) is the delimiter between the conditions. The `destination_out_of_order` condition is defined like this:

```
destination_out_of_order:2:1776.7.
```

This looks for two occurrences of 1776.7 Hz, which is a common SIT tone frequency in the USA (there is a nice introductory article on SIT tones at `http://en.wikipedia.org/wiki/Special_information_tones`). If 1776.7 Hz is heard twice, then the call will fail as `destination out of order`.

When using `monitor_early_media_fail`, only the designated frequencies are detected. All other tones and frequencies are ignored.

Handling no-answer conditions

Handling a no-answer condition is different from busy and other in-band errors. In some cases, the service provider will send back SIP message 480 (NO_ANSWER), whereas others will send a ringing signal (SIP 183) in the early media until the caller decides to hang up. The former scenario is handled automatically by the `bridge` application. The latter can be customized with the use of special timeout variables:

▸ `call_timeout`: Sets the call timeout for all legs when using `bridge`

▸ `originate_timeout`: Sets the call timeout for all legs when using `originate`

▸ `leg_timeout`: Sets a different timeout value for each leg

▸ `originate_continue_on_timeout`: Specifies whether or not the entire `bridge` or `originate` operation should fail if a single call leg times out

By default, each call leg has a timeout of 60 seconds and `bridge`, or `originate`, will stop after any leg times out. The three timeout variables allow you to customize the timeout settings for the various call legs. Use `call_timeout` when using the `bridge` application, and use `originate_timeout` when using the originate API. Use `leg_timeout` if you wish to have a different timeout value for each dial string. In that case, use the `[leg_timeout=###]` square bracket notation for each dial string:

```
<action application="bridge" data="[leg_timeout=10]sofia/internal/
userA@local.pbx.com|[leg_timeout=15]sofia/internal/userB@local.pbx.
com"/>
```

Use `originate_continue_on_timeout` to force `bridge` or `originate` to continue dialing even if one of the endpoints fails with a timeout:

```
<action application="bridge"
data="{originate_continue_on_timeout=true}[leg_timeout=10]
sofia/internal/userA@host|[leg_timeout=15]sofia/internal/
userB@host"/>
```

Keep in mind that by default, a *timeout* (that is, a no answer) will *end* the entire `bridge` or `originate` if you do not set `originate_continue_on_timeout` to `true`.

Another thing to keep in mind is handling cases where you are calling a phone number that has voicemail. For example, if you are trying to implement a type of "find me, follow me" and one of the numbers being called is a mobile phone with voicemail, you need to decide whether you want that phone's voicemail to answer your call. If it does answer, then the `bridge` will be completed. If you do not want the voicemail to answer and end the `bridge` (so that your `bridge` will keep dialing the remaining endpoints), then be sure to set `leg_timeout` to a relatively low value. If the voicemail picks up after 15 seconds, then you may wish to set `leg_timeout=12`. In most cases, you will need to make several test calls to find the best timeout values for your various endpoints.

Using individual bridge calls

In some cases, you may find that it is helpful to make a dial attempt to a single endpoint and then do some processing prior to dialing the next endpoint. In these cases, the pipe-separated list of endpoints will not suffice. However, the FreeSWITCH XML dialplan allows you to do this in another way. Consider this excerpt:

```
<extension name="ring_sequentially">
<condition field="destination_number" expression="^(2001)$">
<action application="set" data="continue_on_fail=true"/>
<action application="set" data="hangup_after_bridge=true"/>
<action application="bridge" data="{ignore_early_media=true}
sofia/internal/userA@local.pbx.com"/>
<action application="log" data="INFO call to userA failed."/>
<action application="bridge" data="{ignore_early_media=true}
sofia/internal/userB@local.pbx.com"/>
<action application="log" data="INFO call to userB failed."/>
</condition>
</extension>
```

Key to this operation are the highlighted lines. In the first of them, we set `continue_on_fail` to `true`. This channel variable tells FreeSWITCH to *keep processing* the actions in the extension even if a `bridge` attempt fails. After each bridge attempt, you can do some processing. Note, however, that we set `hangup_after_bridge` to `true`. This is done so that the dialplan does not keep processing after a successful `bridge` attempt (for example, if the call to `userA` was successful, we would not want to call `userB` after `userA` hung up). You may add as many additional `bridge` endpoints as you need.

See also

▶ The *Ringing multiple endpoints simultaneously* and *Advanced multiple endpoint calling with enterprise originate* recipe in this chapter

Advanced multiple endpoint calling with enterprise originate

You've seen many ways of ringing multiple destinations with many options, but in some cases even this is not good enough. Say you want to call two destinations at once, but each of those two destinations is a separate set of simultaneous or sequential destinations.

For instance, you want to call Bill and Susan at the same time, but Bill prefers that you try his cell first, and then try all of his landlines at the same time. Susan, however, prefers that you call her desk first, then her cell, and finally her home. This is a complicated problem, and the solution to it is called **enterprise originate**. The term "enterprise" is used to indicate an increased level of indirection, dimension, or scale. Basically, you are doing everything the originate syntax has to offer, but you are doing entire originates in parallel in a sort of "super originate".

Getting ready

The first thing you need to do to take advantage of enterprise originate is to fully understand regular originate. *Originate is the term used to indicate making an outbound call*. Although there is an `originate` command that can be used at `fs_cli`, the method by which you mostly use the `originate` command is with the `bridge` dialplan application.

The bridge application versus the originate command

Why do we talk about a regular `originate` when discussing the `bridge` application? Are the `bridge` application and the `originate` command not completely different? No! This is a common misconception. The `bridge` application is used in the dialplan, but it does exactly the same thing that the `originate` command does—it creates a new call leg. In fact, `bridge` and `originate` use exactly the same code in the FreeSWITCH core. The only difference between the two is where they are used. The `originate` command is used in `fs_cli` to create a new call leg. The `bridge` application is used in the dialplan to create a new call to which an existing call leg can be connected or bridged.

You will need to open `conf/dialplan/default.xml` in a text editor or edit a new XML file in the `conf/dialplan/default/` subdirectory.

How to do it...

The usage of enterprise originate is similar to the *ring simultaneously* example, but an alternate delimiter (:_ :) is used:

```
<extension name="enterprise_originate">
<condition field="destination_number" expression="^(2000)$">
<action application="bridge"
    data="{ignore_early_media=true}sofia/internal/
userA@local.pbx.com:_:{myoption=true}sofia/internal/
userB@local.pbx.com"/>
</condition>
</extension>

<extension name="enterprise_originate2">
<condition field="destination_number" expression="^(2001)$">
<action application="bridge"
    data="{ignore_early_media=true}sofia/internal/
userA@local.pbx.com,sofia/internal/
userB@local.pbx.com:_:sofia/internal/
userC@local.pbx.com,sofia/internal/userD@local.pbx.com"/>
</condition>
</extension>
```

How it works...

The entire input string is broken down into smaller strings based on the :_ : symbol.

Each of those smaller strings is fed to the regular originate engine in parallel, and the first channel to answer will be bridged to the caller. Once one endpoint answers, the rest of the calls in the enterprise will be *canceled*.

There's more...

Enterprise originate has a few special aspects to consider when using it to place calls.

Setting variables in enterprise originate

As you know, you can use the {var=val} syntax to define special variables to be set on all the channels produced by originate, and [var=val] to define variables per leg in a call with many simultaneous targets. Enterprise originate uses these as well, but remember that each string separated by the :_: delimiter is its own self-contained instance of originate, so {var=val} becomes local only to that single originate string. If you want to define variables to be set on every channel of every originate, you must define them at the very beginning of the string, using the <var=val> notation. This indicates that you should pass these variables to every leg inside every originate. Consider the following enterprise originate:

```
<action application="bridge" data="<ignore_early_media=true>
{myvar=inner1}[who=userA]sofia/internal/userA@local.pbx.com,
[who=userB]sofia/internal/userB@local.pbx.com:_:{myvar=inner2}
[who=userC]sofia/internal/userC@local.pbx.com, [who=userD]sofia/
internal/userD@local.pbx.com"/>
```

At first glance, this may seem confusing, but when you break it down, you can see what the values of the variables are for each channel. This table shows the values:

Channel	${ignore_early_media}	${myvar}	${who}
userA@local.pbx.com	true	inner1	userA
userB@local.pbx.com	true	inner1	userB
userC@local.pbx.com	true	inner2	userC
userD@local.pbx.com	true	inner2	userD

Once you know which syntax to use, it becomes a simple thing to set the channel variables for individual legs inside originates, or the entire enterprise originate.

Ringback

Unlike the regular originate, **signaling** cannot be passed back from one of the inner originates, because there are too many call paths open to properly handle it. Therefore, when using bridge with enterprise originate, you must define the **ringback** variable if you want to send a ringtone back to the caller.

See also

To learn more about originate and enterprise originate, look at some other examples in this chapter and study the default dialplan distributed with FreeSWITCH. There are several examples of the many things you can do when placing outbound calls found in conf/dialplan/default.xml.

Time-of-day routing

It is common for routing of calls to be different, depending on the time of day or day of the week. The FreeSWITCH XML dialplan has a number of parameters to allow this functionality.

Getting ready

Determine the parameters for your routing. In this example, we will define business hours as Monday through Friday from 8:00 a.m. to 5:00 p.m. Additionally, we will include a day_part variable to reflect morning (midnight to noon), afternoon (noon to 6:00 p.m.), and evening (6:00 p.m. to midnight).

How to do it...

Start at the beginning of your dialplan by following these steps:

1. Add this extension to the beginning of your context:

    ```
    <extension name="Time of day, day of week setup" continue="true">
    <condition wday="2-6" hour="8-16" break="never">
    <action application="set" data="office_status=open"
        inline="true"/>
    <anti-action application="set"
        data="office_status=closed" inline="true"/>
    </condition>
    <condition hour="0-11" break="never">
    <action application="set" data="day_part=morning"
        inline="true"/>
    </condition>
    <condition hour="12-17" break="never">
    <action application="set" data="day_part=afternoon"
        inline="true"/>
    </condition>
    <condition hour="18-23" break="never">
    <action application="set" data="day_part=evening"
        inline="true"/>
    </condition>
    </extension>
    ```

2. Later in your dialplan, you can use the `office_status` and `day_part` variables. The `office_status` variable will contain either "open" or "closed", and `day_part` will contain "morning", "afternoon", or "evening". A typical usage would be to play different greetings to the caller, depending on whether or not the office is open. Add these dialplan extensions, which will accomplish the task:

```xml
<extension name="tod route, 5001_X">
<condition field="destination_number" expression="^(5001)$">
<action application="execute_extension"
    data="5001_${office_status}"/>
</condition>
</extension>
<extension name="office is open">
<condition field="destination_number"
    expression="^(5001_open)$">
<action application="answer"/>
<action application="sleep" data="1000"/>
<action application="playback" data="ivr/ivr-
      good_${day_part}.wav"/>
<action application="sleep" data="500"/>
<!-- play IVR for office open -->
</condition>
</extension>
<extension name="office is closed">
<condition field="destination_number"
    expression="^(5001_closed)$">
<action application="answer"/>
<action application="sleep" data="1000"/>
<action application="playback" data="ivr/ivr-
      good_${day_part}.wav"/>
<action application="sleep" data="500"/>
<!-- play IVR for office closed -->
</condition>
</extension>
```

3. Save your XML file and issue the `reloadxml` command at `fs_cli`.

How it works...

The Time of day, day of week setup extension defines two channel variables, namely office_status and day_part. Note the use of inline="true" in our set applications. These allow immediate use of the channel variables in later dialplan condition statements. Every call that hits this dialplan context will now have these two channel variables set (they will also show up in CDR records if you need them). You may have also noticed continue="true" in the extension tag and break="never" in the condition tags. These tell the dialplan parser to keep looking for more matches when it would otherwise stop doing so. For example, without continue="true", when the dialplan matches one of the conditions in the Time of day, day of week setup extension, it stops looking at any more extensions in the dialplan. In a similar way, the break="never" attribute tells the parser to keep looking for more conditions to match within the current extension (by default, when the parser hits a failed condition, it stops processing any more conditions within the current extension).

 A detailed discussion of dialplan processing can be found in Packt Publishing's *FreeSWITCH 1.2* book.

Our sample extension number is 5001. Note the action it takes:

```
<action application="execute_extension"
data="5001_${office_status}"/>
```

This sends the call back through the dialplan looking for a destination_number of 5001_open or 5001_closed. We have defined these destinations with the "office is open" and "office is closed" extensions respectively. Now we can play different greetings to the caller: one when the office is open and a different one when the office is closed. As a nice touch, for all calls, we play a sound file that says "Good morning", "Good afternoon", or "Good evening", depending on what the value in the day_part channel variable is.

The execute_extension and transfer dialplan applications

These two applications tell FreeSWITCH to execute another part of the dialplan. The primary difference is that execute_extension will return after executing another portion of the dialplan, whereas transfer will send control to the target extension. In programming parlance, execute_extension is like a gosub command and transfer is like a goto command. The former comes back, but the latter does not.

There's more...

You may be wondering why we did not simply use a `condition` to test `office_status` for the `open` value, and then use `action` tags for "office open" and `anti-action` tags for "office closed". There is nothing preventing us from doing this. However, what if you need to have an office status other than "open" or "closed"? For example, what if you have an office that needs to play a completely different greeting during lunch time? This is difficult to accomplish with only `anti-action` tags, but with our example, it is very simple. Let's make it a bit more challenging by adding a lunch period that runs from 11:30 a.m. to 12:30 p.m. We cannot use `hour="11.5-12.5"`, but we do have another value we can test — `time-of-day`. This parameter lets us define periods in the day at a granularity of minutes, or even seconds. The value range can be 00:00 through 23:59 or 00:00:00 through 23:59:59. Consider this new `Time of day, day of week setup` snippet:

```
<extension name="Time of day, day of week setup" continue="true">
<condition wday="2-6" hour="8-16 break="never">
<action application="set" data="office_status=open"
    inline="true"/>
<anti-action application="set" data="office_status=closed"
    inline="true"/>
</condition>
<condition wday="2-6" time-of-day="11:30-12:30" break="never">
<action application="set" data="office_status=lunch"
    inline="true"/>
</condition>
    . . . .
```

Notice that we need to explicitly define the weekend, since we cannot rely on a simple Boolean "open" or "closed" condition. However, we now have a new `office_status` of "lunch" available to us. We define an additional extension to handle this case:

```
<extension name="office is at lunch">
<condition field="destination_number"
  expression="^(5001_lunch)$">
```

Add the specific dialplan actions for handling calls during the office's lunch hour, and you are done. You can add as many new office statuses as you need.

See also

- ▶ Refer to the XML dialplan page at `http://freeswitch.org/confluence/display/FREESWITCH/XML+Dialplan` for more information on the usage of the `break`, `continue`, and `inline` attributes.

Manipulating SIP To: headers on registered endpoints to reflect DID numbers

Sometimes, when routing calls to endpoints that are registered with your system, you would want to utilize custom SIP To: headers. For example, if you are routing DIDs to a PBX or switch, the device you are sending the call to might expect the phone number you wish to reach in the To: SIP header. However, the customer or PBX may have only a single registration to your service that represents multiple DIDs that need to be sent to them.

By default, no flags exist for changing the To: header to match the DID when calling a registered endpoint. Since the registration to your server is typically done via a generic username that is not related to the DID, you must program your dialplan to retrieve a user's registration information and parse out the username portion of the To: header, replacing it with your own. Care must be taken to replace only the username portion and keep the remaining parameters (after @) intact, especially if the NAT traversal is expected to continue operating.

Getting ready

Be sure that you have your DIDs and users configured. In this example, we will use testuser as the username, with a phone number of 4158867999, and our domain will be my.phoneco.test.

How to do it...

Create a dialplan extension specifically for handling calls to the DID number, and use some regular expression syntax to parse out the information. Here is an example:

```
<extension name="call_4158867999">
<condition field="destination_number"
  expression="^\+?1?4158867999$"/>
<condition field="${sofia_contact(testuser@local.pbx.com)}"
  expression="^[^\@]+(.*)">
<action application="bridge"
    data="sofia/external/4158867999$1"/>
</condition>
</extension>
```

How it works...

You would typically send calls to `testuser` using the `bridge` command with an argument of `user/testuser`. In this scenario, however, you would wish to call the testuser's registered endpoint and replace `testuser` with a phone number, which is `4158867999` in our example. To do this, you must retrieve the testuser's current dial string and remove the username, replacing it with the DID number.

In this example, we leverage the `sofia_contact` API and some regular expression magic. The first `condition` simply matches the user's DID phone number. We only want to act if the destination number is `4158867999`. The interesting stuff happens in the second `condition`. The field is `${sofia_contact(testuser@local.pbx.com)}`. By wrapping an API call in `${}`, the dialplan literally executes the API and uses the result as the field value. If we go to `fs_cli` and type `sofia_contacttestuser@local.pbx.com`, we get the result, which is something like this:

```
sofia/external/johndoe@12.34.56.7;fs_nat=yes
```

The `^[^\@]+(.*)` regular expression pattern is applied against this value. The result is that everything after, and including, the @ sign is placed in the `$1` variable. In our example, `$1` contains `@12.13.56.7;fs_nat=yes`. Finally, we execute `bridge` with the `sofia/external/4158867999$1` dial string. With `$1` expanded, our destination is as follows:

```
sofia/external/4158867999@12.34.56.7;fs_nat=yes
```

We have successfully replaced `testuser` with `4158867999`, while preserving the necessary IP address and parameters for contacting the server, and sent the call to the proper destination.

2
Connecting Telephones and Service Providers

In this chapter, we will cover these recipes:

- ▶ Configuring an SIP phone to register with FreeSWITCH
- ▶ Connecting to Skype
- ▶ Configuring an SIP gateway
- ▶ Codec configuration

Introduction

As its name implies, FreeSWITCH will "switch" or "connect" various endpoints together. A part of this switching involves making semi-permanent connections to individual telephones or telephone service providers. Service providers are usually telephone companies (telcos), or **Internet Telephony Service Providers** (**ITSP**). Continue reading to learn more about the many ways in which FreeSWITCH can connect your telephone to the world.

The recipes in this chapter will delve into the various ways to connect FreeSWITCH to telephones and service providers. FreeSWITCH can also connect to Skype using `mod_skypopen`. The last recipe is for advanced users, and discusses the subject of codec negotiation.

Configuring an SIP phone to register with FreeSWITCH

SIP phones, or any SIP devices with the ability to register, are essential in most FreeSWITCH installations for allowing users to communicate with each other. A registration is when a phone or other device informs FreeSWITCH that it is active, and provides information (such as an IP address and port) on how to reach the phone across the network or Internet. FreeSWITCH stores this information for later use to contact that phone.

In this recipe, you will be registering a phone to FreeSWITCH. You will need to enter your credentials into your phone as well as into FreeSWITCH itself (both sides must match).

Getting ready

Ensure that the `mod_sofia` module is already compiled and loaded (Sofia is the SIP stack). You may also want to know on which IP address your registrations are being accepted.

Follow these steps:

1. Launch the FreeSWITCH command line interface.

2. To view the current ports and IPs that you are listening on, type

   sofia status.

3. Review the output, specifically the lines listed as `internal` and `external`, as follows:

external is the profile that is generally used to receive incoming calls, and to send outgoing traffic to our provider (from which we'll go to the destinations), and **internal** is the profile to which your internal phones and the phones of your users are registered to. Profiles are associated with a specific port and IP address. In this example, internal is associated with the 192.168.1.124 IP, implicitly port 5060. In essence, this means that the registrations that go to and fro from FreeSWITCH, for the internal profile, should occur on that address and port.

How to do it...

From an SIP standpoint, a real hardware phone or a software that runs on your desktop (or your cellphone) is exactly the same.

So, first of all, if you don't own a hardware SIP phone, install an SIP softphone. In this recipe we'll use X-Lite from CounterPath (`http://www.counterpath.com/x-lite`), but you can use whatever other SIP softphone suits your operating environment. For a list of softphones, check the FreeSWITCH documentation at `https://freeswitch.org/confluence/display/FREESWITCH/Softphones`.

If you own a hardware SIP phone, you'll find that configuring it will require the same information.

The following steps will show you how to configure an SIP phone:

1. Decide on a new username and password that you wish to register with.
2. Create and edit the *new* file called `directory/default/USERNAME.xml` in the FreeSWITCH configuration directory (usually `/usr/local/freeswitch/conf` or `/etc/freeswitch`). Replace USERNAME with a name or an extension number that is not yet there (such as `2000`).
3. Add the following content to the file and save it:

```
<include>
  <user id="2000">
    <params>
      <param name="password" value="PASS"/>
    </params>
  </user>
</include>
```

Replace `2000` and `PASS` in the code with a username and password of your choosing.

4. Load the FreeSWITCH CLI being used.
5. Reload the in-memory configuration in FreeSWITCH's CLI by typing:

```
reloadxml
```

You will now be able to configure your softphone or device to register with FreeSWITCH. To do this, set the username as 2000 as shown, and keep the password as PASS within your softphone or device. Set to register to the IP address (and port, if not the standard 5060) that you identified earlier.

For example, if you created the username 2000 and password PASS, you would enter the following into your softphone:

How it works...

Let's see what you've done.

You've defined an SIP **Username** in the `<user id="">` field. This username is used for the authentication of the SIP packets.

You've added this option in a file within the `directory/default/` folder, which includes it as part of the default directory domain. This domain, by default, is your server's domain name (probably an IP address on your system).

There's more...

The SIP registration shown earlier was extremely basic. It doesn't specify a **context** for their calls to be placed in, it doesn't set the **Caller ID** for the device/user, and it doesn't add any extra variables to the account. Let's talk about these options, as they are common additions to any registration and directory entry. Also, if you don't specify the context, FreeSWITCH will put the incoming call in the "public" context by default, and that context will just drop the call if not otherwise customized.

Context

Calls received by FreeSWITCH will be directed, by default, to the context that is associated with the port and IP that a call comes in on. For example, calls received on port 5060 that are authenticated are assumed to be from an "in-house phone", and get to use the `default` context. If, for some reason, you want to override the default for a particular device with a special context selection, you can do so by adding an additional variable.

Customizing caller ID

You want user `2000` to have a specific caller ID using the previous example. You can make this happen by defining a variable within the user's definition.

The example sets the Caller ID Name to `Mary Sue` and the Caller ID Number to `2000`. Note that, if you choose to, you can further override this setting within the dialplan:

```
<include>
  <user id="2000">
    <params>
      <param name="password" value="PASS"/>
    </params>
    <variables>
      <variable name="effective_caller_id_name" value="Mary Sue"/>
      <variable name="effective_caller_id_number" value="2000"/>
      <variable name="user_context" value="default"/>
    </variables>
  </user>
</include>
```

```
"2000.xml" 13L, 341C written                                    9,50        All
```

In the user definition, adding the `user_context` variable will route all the calls from this device to the context named as "default" NB: actually, "default" is a name (of an interesting context: try calling 5000). The default context is named "public").

See also

▸ *Chapter 4, External Control*

Connecting to Skype

Skype™ is often the first experience of VoIP for many users, and many of them continue to use it as a voice, chat, and video medium to communicate. While many of its features are available in many other kinds of network and technologies (both proprietary and open), the sheer power of ease of use and installed base makes Skype a mainstay of communication.

FreeSWITCH can be made to fully interact with the Skype Network, for both incoming and outgoing voice calls, and for chat messaging.

Getting ready

The `mod_skypopen` module is already compiled for Windows users when using the Visual Studio 2008/2010 solution files with the FreeSWITCH source code. Linux users will need to enable `mod_skypopen` in their FreeSWITCH installation (`mod_skypopen` is not available for Mac and *BSDs at the moment). To do this, follow these steps:

1. Install and configure all the prerequisites ("required packages") for your Linux distribution as per `http://freeswitch.org/confluence/display/` `FREESWITCH/mod_skypopen`. There are a lot of prerequisites, but they can all be installed in a big scoop by copying and pasting from the link.

2. Open `modules.conf` in the FreeSWITCH source directory, and remove the comment on the `#endpoints/mod_skypopen` line. Save the file and exit.

3. Compile `mod_skypopen` using the following command:

    ```
    make mod_skypopen -install
    ```

4. Use the interactive installer to download and setup all the ancillary software needed by `mod_skypopen`. The interactive installer will ask you questions, and then will create all needed directories, configuration files, scripts, and more. Again, if in doubt, please check the documentation page. Remember to use a specific Skype account for `mod_skypopen`; for example, a skypename that is not used by you or anybody else. You can create a new skypename (account) from your desktop Skype client after signing out from your own account. We used the "gmaruzz5" Skype account for `mod_skypopen`.

5. Execute the script that has been created by the interactive installer (sh /usr/ local/freeswitch/skypopen/skype-clients-startup-dir/start_ skype_clients.sh). It will start the X servers and Skype clients that are needed by mod_skypopen.

6. If you want to have mod_skypopen loaded automatically each time you start FreeSWITCH, then edit conf/autoload_configs/modules.conf.xml and uncomment the following line:

    ```
    <!-- <load module="mod_skypopen"/> -->
    ```

7. Save the file and exit.

8. If you do not load mod_skypopen automatically, then simply load it using the following command from fs_cli:

    ```
    load mod_skypopen
    ```

Once mod_skypopen is loaded, you are ready to make and receive Skype audio calls and chat.

How to do it...

The first thing to do is to become familiar with the sk command—sk being short for Skypopen.

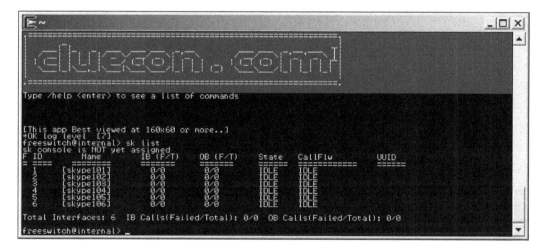

1. At fs_cli, type sk list, and press *Enter*. You will see a list of all the skypopen interfaces (or channels) that you have configured, and their status. In this case, we configured (at installation time via the interactive installer) 6 skypopen interfaces. This FreeSWITCH server can manage at most 6 concurrent Skype voice calls.

2. We've seen that we have 6 interfaces, all of them idle (not serving a call). We can test the full functionality making an outbound Skype voice call from the FreeSWITCH server to our own "usual" desktop. If your personal Skype account is "gmaruzz4", then from fs_cli, type:

```
freeswitch@internal> bgapi originate skypopen/ANY/gmaruzz4 5000
```

3. FreeSWITCH will *originate* in background (bgapi) a call to "gmaruzz4" using "ANY". This is one of the skypopen available interfaces. Then, when answered, it will connect that call to the 5000 extension of the default dialplan.

4. You will receive a call on your personal Skype account ("gmaruzz4") from the account that you assigned to mod_skypopen with the interactive installer (in our case, "gmaruzz5"). When you answer the call, you'll be connected to the standard FreeSWITCH IVR (at extension 5000). You can then ask your desktop Skype client to show you the dialpad numeric keyboard (from the call menu), and navigate the voice tree.

How it works...

Skypopen allows FreeSWITCH to use local skype clients, running in the background on Xvfb "fake" Xservers as endpoints. Mod_skypopen exchanges commands with the local Skype client (as it would similarly to a modem), and redirect the audio streams from and to the local Skype client and FreeSWITCH. For example, mod_skypopen acts as a "remote control" of the local Skype client. From FreeSWITCH's standpoint, mod_skypopen is just another endpoint, and can be used for both, receiving the inbound Skype calls and for originating the outbound voice calls (and messaging chats). If the Skype account used by mod_skypopen has "credits", then you can also send an SMS (cellphone text message), and originate national and international PSTN and mobile calls. From fs_cli, type the following (substitute your cellphone number to the dummy):

```
freeswitch@internal> bgapi originate skypopen/ANY/+12125551212 5000
```

There's more...

`Mod_skypopen` can also receive calls (and chats).

From your own desktop (or smartphone) Skype account, call the account that you have assigned to `mod_skypopen`, with the interactive installer (in our case, "gmaruzz5").

FreeSWITCH will receive your call and will route it to the standard IVR (extension 5000). In our example, after a while, we type 3 on the Skype desktop dialpad, and we'll get transferred to music on hold.

This 5000 destination was set up at installation.

Incoming Skype calls connected to an SIP softphone

To change the destination of inbound Skype calls, you can edit the `/usr/local/freeswitch/conf/autoload_configs/skypopen.conf.xml` file (it was created by the interactive installer). Look at the first line and you'll see:

```
<param name="destination" value="5000"/>
```

You can change this value to be an arbitrary extension in the dialplan.

Let's change it to "2000". From `fs_cli`, enter the `reload mod_skypopen` command to make skypopen aware of the modified configuration.

However, an extension "2000" does not yet exist in the dialplan. Let's add it so that it will connect the incoming call to the softphone that we've registered in the previous paragraph. Edit the `/usr/local/freeswitch/conf/dialplan/default.xml` file, and find "Local Extension" in it. Modify the line that follows "Local Extension" from:

```
<condition field="destination_number"
expression="^(10[01][0-9])$">
```

To:

```
<condition field="destination_number"
expression="^(20[01][0-9])$">
```

For example, change the first 1 and make it 2.

Then, from `fs_cli`, issue the `reloadxml` command (so that it makes the modification to the dialplan). Now, if you call "gmaruzz5" from your desktop Skype, you will be connected to your SIP softphone!

Incoming SIP calls connected to Skype destinations

Let's modify the dialplan again. This time, we will add an extension that will allow the calls to Skype destinations. We will then go to that extension from our SIP softphone, reopen the dialplan file that we just modified, and at the beginning of it find the line:

```
<context name="default">
```

Immediately below it, insert the following lines:

```
<!-- dial echo123 via skypopen using ANY interface to go out -->
<extension name="skypopen">
  <condition field="destination_number" expression="^2909$">
    <action application="bridge" data="skypopen/ANY/echo123"/>
  </condition>
</extension>
```

Then, again from `fs_cli`, issue the `reloadxml` command. Now if you call "2909" from your SIP softphone, you will be connected to the Skype test ("echo123" Skype account).

Configuring an SIP gateway

Configuring an SIP gateway allows you to connect with outside carriers or other SIP machines. You can connect with other FreeSWITCH or Asterisk boxes, or upstream carrier SIP trunks.

SIP gateways have many, many options—too many to list here; so, we'll review just a few.

Getting ready

First, you'll need to gather some information about the remote server to which you want to connect. The list generally includes the following:

- The IP address (or hostname) and a port (the standard port is 5060) of the server that to which you are being connected
- The username and password (if any)
- How the carrier/gateway expects Caller ID to be handled (which SIP header the Caller ID should be placed in)
- Whether registration is required

You'll also need to know the phone number format that your carrier expects when you send calls to them, and in which format they'll send calls to you.

Finally, you'll need to decide which of your existing SIP profiles you want to tie this gateway to. All gateways must be associated with an SIP profile (port and IP address). Note that, in most cases, a gateway can be utilized on multiple SIP profiles if desired.

 Some carriers use SIP registration to figure out how to send calls to you, while other carriers map IP and port addresses permanently to deliver calls to you. Some carriers also allow DNS-based records to be used. You should find out what your provider utilizes. You may be allowed to tweak these options within the provider's configuration interface.

How to do it...

Gateways are associated with SIP profiles, because FreeSWITCH needs to know which IP and port to send traffic to and from in relation to the carrier.

First, you'll need to add a gateway to your SIP profile. Let's assume you're using the default FreeSWITCH configuration. In this case, we'll create a gateway that is attached to the default external profile.

1. Create a file in the `conf/sip_profiles/external/` directory named after your gateway (that is, `cheap_tel.xml`).

2. Add the following content (note that even if you are not registering, a username and a password are required) and replace the highlighted items with your own provider data:

```
<include>
  <gateway name="providerA">
    <param name="realm" value="sip.2600hz.com"/>
    <param name="username" value="darren"/>
    <param name="password" value="test"/>
    <param name="register" value="true"/>
  </gateway>
</include>
```

3. You will access the gateway by using the `bridge` application with `sofia/gateway/providerA/number`, such as `sofia/gateway/providerA/4158867999`. You can do this in any dialplan that you are using. In this example, edit your dialplan (typically, the default dialplan in `conf/dialplan/default.xml`) and add the following code to utilize the gateway:

```
<extension name="dial-10-digit-numbers">
  <condition field="destination_number"
             expression="^(\d{10})$">
```

```
<action application="bridge"
        data="sofia/gateway/providerA/$1"/>
    </condition>
</extension>
```

4. Issue a `reloadxml` command in your FreeSWITCH CLI after making the mentioned changes.

5. Issue `sofia profile external rescan` to instruct FreeSWITCH to find any new gateways or settings on the profile `external`.

Sofia profile rescan versus reload

When making changes to your SIP configuration files, you will have to tell FreeSWITCH's SIP module ("Sofia") that you want these changes to take effect. Simply reloading the XML configuration in memory (`reloadxml`) does not force sofia to apply it. Instead, you will need to tell sofia profile to `rescan` or `reload`.

The `reload` option will completely stop the sofia profile, dropping all calls in progress, and will then restart the profile with the new changes that are applied. The `rescan` option is much less intrusive. Instead of stopping the profile altogether, it simply looks for the changes made in the XML configuration, and selectively applies them. Changes to a gateway only require a `rescan`. However, changes made to the sofia profile parameters require `reload`.

How it works...

In step 2, you defined in the profile a very basic gateway containing a gateway name, a server name, a username, and a password. In step 3, you added to the dialplan a condition that matched destination numbers that were composed of no more and no less than 10 digits (long-distance in the USA numbering plan), and bridged the calls to such numbers using your new gateway.

Note that in the `bridge` application in step 3, you utilized `$1` as a variable that contains, which was captured before by the parenthesis in your regular expression to pass along the number that was dialed.

Step 4 and step 5 tell FreeSWITCH to load your new profile into memory and activate it.

There's more...

Connecting to a provider is usually the first step in configuring outbound calls. The following sections provide additional information on how to make your FreeSWITCH gateways more effective.

Adding prefixes to dial strings

You can add **prefixes** to the `bridge` dial strings in multiple ways. In the simplest form, you might want to add a country or area code to the beginning of a number. In the example, if you modify the `bridge` string from `sofia/gateway/providerA/$1` to `sofia/gateway/providerA/+1$1`, your calls will now be completed with a prefix of +1 (the USA international prefix) in front of the 10-digit (the USA long-distance format) number. The complete number, made by the international prefix, then the long distance prefix, then the number, is commonly referred to as **E.164** format.

Another common strategy is to add an *account code* or *customer code* to the beginning of a number. To do this, you can add a prefix that is based on a channel variable. In this scenario, let's say you have a customer with the account code 38234, and a customer with the account code 93289. Each customer makes calls from a specific IP address. You might have an XML dialplan that looks like this:

```
<extension name="check_customer_1">
  <condition field="network_addr" expression="^2\.3\.4\.5$">
    <action application="set" data="accountcode=38234"
           inline="true"/>
  </condition>
</extension>


<extension name="check_customer_2">
  <condition field="network_addr" expression="9\.8\.7\.6$">
    <action application="set" data="accountcode=93289"
           inline="true"/>
  </condition>
</extension>


<extension name="dial-10-digit-numbers">
  <condition field="accountcode" expression="^.$"/>
  <condition field="destination_number" expression="^(\d{10})$">
    <action application="bridge"
           data="sofia/gateway/providerA/${accountcode}$1"/>
  </condition>
</extension>
```

In the example, we first set the appropriate `$accountcode` variable (inline) during the dialplan, processing to identify the client (if you set a variable in dialplan and want to check it during the same dialplan "pass", then you must declare it as "inline" or it will be available only in the next "pass".)

Then, if the `accountcode` variable is empty, we do nothing and exit the extension (implicitly hanging up), because the condition that matches the empty `accountcode` variable - "`^.$`" - is closed at the end, and does not contain any actions. So, exit. If that condition is not matched, we check the next one. So, we bridge to the provider only if the `accountcode` variable is set (if it was empty we would not even reach the line that contains the `bridge` action), and utilize `accountcode` as a prefix in the dial string (see the bold portion of the `bridge` command).

Monitoring gateways

There are many additional parameters available on your gateway profile. One such parameter is the OPTIONS **ping** setting. This tells FreeSWITCH to send periodically (to "ping") the OPTIONS SIP packets to the gateway, and check the answer to ensure it's up. This is useful in ensuring that if the gateway is marked "down", you do not hangup while trying to reach it, and can instead do error handling and/or move on to another gateway/carrier.

To implement OPTIONS pings, add this parameter to your gateway definition (step 2 in *How to do it*):

```
<param name="ping" value="25"/>
```

This will ping the gateway every 25 seconds to ensure it's up. If the gateway is marked "down" because it has not responded to pings, FreeSWITCH will continue sending OPTIONS pings at the specified interval, and will mark it "up" as soon as it responds.

Codec configuration

Codec configuration is very versatile in FreeSWITCH. In IP telephony, there are several differing scenarios for negotiating and choosing codecs. To meet the varying demands, FreeSWITCH has several configurable modes of operation as well as runtime variables that can influence how codec negotiation takes place. Typically, the goal should be to reduce **transcoding** or **resampling** as much as possible. Transcoding is the case where two sides of the call have different codecs and audio flowing in either direction, which has to be completely decoded and re-encoded to the opposite channel's format. Resampling is similar, but it is required when each side of the call is running at a different sample rate, and the audio has to be "resampled" to the correct rate. One or both of these can be necessary, depending on where you direct your calls to, and how you have your codec configuration set.

Getting ready

The biggest decision to make upfront is **late-negotiation** or **early-negotiation**. This setting tells FreeSWITCH to either validate the codec before the channel even hits the dialplan, or wait until the moment where media is absolutely necessary to perform codec negotiation. This gives you a chance to decide on a codec from your dialplan logic, or even from the result of a separate outgoing call that you intend to bridge.

With early-negotiation, there is not much you can do to control the codec behavior of inbound calls. So, for this recipe, we will work with late-negotiation. To prepare, follow these steps:

1. Open internal sofia profile `conf/sip_profiles/internal.xml` in a text editor and look for this line:

   ```
   <!--<param name="inbound-late-negotiation"
   value="true"/>-->
   ```

2. Uncomment this parameter to enable late-negotiation for all calls.

3. Save the file and exit. At `fs_cli`, issue the `reloadxml` command, and then issue the sofia profile internal restart command.

You are now ready to experiment with codec late-negotiation.

How to do it...

Late-negotiation gives you the possibility of trying your best to ensure that the two legs of a call use the same codec (for example, no transcoding occurs). You can activate this possibility by setting the `inherit_codec` variable. This variable will see the list of codecs proposed by leg A (delaying the decision on which one to accept), then it will arrange the order of that list (to suit its own preferences), and then it will propose it to leg B. Leg B will choose, and the FS will choose the same for leg A:

1. Add the following extension to your dialplan. Create `conf/dialplan/default/01_codec_negotiation.xml` and add these lines:

   ```xml
   <include>
     <extension name="example">
       <condition field="destination_number" expression="^1234$">
         <action application="set" data="inherit_codec=true"/>
         <action application="bridge"
                 data="sofia/internal/1234@cluecon.com"/>
       </condition>
     </extension>
   </include>
   ```

2. Save the file and exit. At `fs_cli`, issue the `reloadxml` command.

How it works...

First, let's see how to explicitly dictate codec negotiation. Once the late-negotiation parameter is set, you can set a special channel variable called `absolute_codec_string`. This variable is in the same format as all other codec parameters inside FreeSWITCH, and contains a comma-separated list of codec names with modifiers to choose the rate or interval, such as `G729`, `PCMU@30i`, or `speex@16000h`. The `@i` sets the interval (milliseconds of audio per packet) and the `@h` sets the hertz (sampling rate) of the codec. A simple dialplan that sets `absolute_codec_string`, and then places an outbound call can demonstrate how to explicitly choose a codec using late-negotiation:

```
<extension name="example">
  <condition field="destination_number" expression="^888$">
    <action application="set"
            data="absolute_codec_string=PCMU@30i"/>
    <action application="conference" data="888@default"/>
  </condition>
</extension>
```

Let's take it a step further. Say you are placing an outbound call to one or more servers and you want to avoid transcoding, but you don't know what codecs that outbound call will be offered, and it is too late at this point to set `absolute_codec_string`. The solution is to use `inherit_codec`. This variable, when set to `true`, tells FreeSWITCH to automatically set `absolute_codec_string` to the value of the codec that was negotiated by the outbound leg in the case of a bridged call. This way, you can allow the outbound call to negotiate a codec, and then pass that decided value back to the inbound leg before media is established. This will then force the inbound leg to request the same codec as the outbound leg, and try to eliminate transcoding.

When calling the example extension, the call hits the XML dialplan and executes the instructions contained in the `action` tags. First, the `inherit_codec` variable is set to `true`, and then the call is bridged to `1234@cluecon.com` over SIP. Because we previously enabled the `inbound-late-negotiation` parameter in the profile, the codec has not yet been chosen for the inbound leg. The outbound leg then proceeds to connect to `http://cluecon.com/`, where a codec will be chosen when the far-end answers or establishes media. At this point, the FreeSWITCH call origination engine will take the codec from the outbound leg and set it as `absolute_codec_string` on the inbound leg. Next, the media indication is passed across that will prompt the inbound leg to negotiate media and offer the same codec as the outbound leg.

There's more...

You can also limit the codecs that you offer to the outbound leg with another special variable called `ep_codec_string`. The `ep_codec_string` variable contains the list of codecs offered by the calling endpoint (**A leg**). This variable is the same one used by the `inherit_codec` behavior, and can be used on an inbound call to make sure that you only offer codecs on the outbound leg that were initially offered to the inbound leg. Here is the previous example with this extra functionality enabled:

```
<extension name="example">
  <condition field="destination_number" expression="^1234$">
    <action application="set" data="inherit_codec=true"/>
    <action application="export"
     data="nolocal:absolute_codec_string=${ep_codec_string}"/>
    <action application="bridge"
            data="sofia/internal/1234@cluecon.com"/>
  </condition>
</extension>
```

The `export` application sets the desired variable on the inbound leg just as the `set` application, but marks it to be copied to (that is, "exported" to) any outbound call legs generated by the channel on which it is set. The `nolocal:` syntax prevents the variable from applying to the channel on which it was set, but still copies it to any outbound legs. So in this case, we use export to set `nolocal:absolute_codec_string` to the current value of `ep_codec_string` for any outbound calls. This means that when we bridge to `1234@cluecon.com`, our `absolute_codec_string` will be set to exactly what codecs the inbound leg was offered.

Further information and techniques for codec negotiation in FreeSWITCH:
`http://freeswitch.org/confluence/display/FREESWITCH/Codec+Negotiation`

Avoiding codec negotiation altogether

It's also possible to route your calls to a script or some other application that does not require media, and uses logic to influence the `absolute_codec_string` in similar ways to what was demonstrated earlier. If you want to try to be completely uninvolved with the codec negotiation, you can try setting the `bypass_media` variable to `true` before you call the `bridge` application. FreeSWITCH will present the inbound SDP to the outbound leg and vice-versa, completely eliminating FreeSWITCH from the media path, but still keeping it in the signaling path. This, however, does not work well under NAT conditions.

3

Processing Call Detail Records

In this chapter, we will cover these recipes:

- ▶ Using CSV CDRs
- ▶ Using XML CDRs
- ▶ Inserting CDRs into a backend database
- ▶ Using a web server to handle XML CDRs
- ▶ Using the event socket to handle CDRs
- ▶ Directly inputting CDRs into various databases in real time

Introduction

Call detail records (**CDR**) are an important part of the accounting process of any phone system. They are also an invaluable resource for troubleshooting. FreeSWITCH provides several different methods for the generation of CDRs. The most common method is to create plain-text **comma-separated values** (**CSV**) files. Each line in a CSV file represents one phone call or, more accurately, one call leg. (A call is often made by two "legs": one leg, called **leg A**, is incoming from the caller to FreeSWITCH; the other leg, called **leg B**, is outbound from FreeSWITCH to the callee. FreeSWITCH "bridges" the audio of the two call legs so that the caller and callee can talk to each other.) There are other options for processing CDRs, most notably using mod_xml_cdr to store more detailed information about calls, as well as using the event socket to process CDR information.

Using CSV CDRs

It is a simple thing to store CDRs in CSV format. This recipe describes the steps necessary to store call records in plain-text CSV files.

Getting ready

In the default configuration, `mod_cdr_csv` is compiled and enabled by default. CDR data is stored in the `log/cdr-csv/` directory. To review the options available, open the `conf/autoload_configs/cdr_csv.conf.xml` file. Here are the parameters available in the settings section:

```
<settings>
    <!-- 'cdr-csv' will always be appended to log-base -->
    <!--<param name="log-base" value="/var/log"/>-->
    <param name="default-template" value="example"/>
    <!-- This is like the info app but after the call
     is hung up -->
    <!--<param name="debug" value="true"/>-->
    <param name="rotate-on-hup" value="true"/>
    <!-- may be a b or ab -->
    <param name="legs" value="a"/>
    <!-- Only log in Master.csv -->
    <!-- <param name="master-file-only" value="true"/> -->
</settings>
```

We will review some of these options in the following section.

How to do it...

The easiest way to see a new CDR is to use a utility such as `cat` in Linux/Unix or `type` in Windows. Alternatively, if you are in a Linux/Unix environment, you can use the `tail` utility to see in real time what is appended to the end of a text file. (Windows does not ship with a `tail` utility, but there are free and open source options available. Also, take a look at Cygwin, a complete and open source Unix/Linux-like environment for Windows.)

Here are the steps you can follow in a Linux/Unix environment:

1. Change the directory to `/usr/local/freeswitch/log/cdr-csv/`.
2. Execute `tail -f Master.csv` to display new CDR entries as they are written.
3. Make a test call, perhaps from one phone to another.
4. Hang up the test call and note the new CDR, which is appended to `Master.csv`.
5. Press *Ctrl + C* to exit the `tail` command.

Here are two sample CDRs. The first one is from a call made by gmaruzz4 to 2000, and the second is from a call made by 2000 to 2909:

```
"Giovanni Maruzzelli","gmaruzz4","2000","defau
lt","2015-03-15 03:47:06","2015-03-15 03:47:10","2015-03-15
03:47:20","14","10","NORMAL_CLEARING","55232714-a5a2-4278-b53d-
7f266d73314e","4ac1721f-09ed-4c45-9436-36f8618003ca","","L16","L16"

"2000","2000","2909","default","2015-03-15 04:06:34","2015-03-15
04:06:36","2015-03-15 04:07:27","53","51","NORMAL_CLEARING","4e048b90-
78e9-47b8-8ee0-db3c26b09a92","769c594f-0c67-4daf-8a59-
e0c641d8709c","","opus","opus"
```

How it works...

By watching the `Master.csv` file, we can observe new CDRs being written to the disk. While not particularly useful in a production system, doing this helps you learn about CDRs and the information they contain. Furthermore, it is a simple troubleshooting tool you can use down the road.

There's more...

There are a number of things to keep in mind when using CSV CDRs. The following sections will help you make the best use of them.

File names and locations

If you perform a directory listing of `log/cdr-csv`, you may see a number of files in addition to `Master.csv`. For example, using the default configuration, if you make a call from 1001 to 1007, you will see a file named `1001.csv`. The reason is that those two users are defined by the files in the `/usr/local/freeswitch/conf/directory/default/*.xml` directory, and inside their definition, we can find an "accountcode" parameter. Each "accountcode" will generate its own `.csv` file, which contains all the CDRs made by users sharing that accountcode. In the default configuration, each directory user is defined with its own accountcode (for example, user 1001 has the accountcode 1001), but you can set up multiple users sharing the same accountcode (for example, both 1001 and 1007 have the accountcode "sales"). These files are in addition to `Master`, and are a feature that can be disabled in `conf/autoload_configs/cdr_csv.conf.xml` by setting this parameter:

```
<param name="master-file-only" value="true"/>
```

You may see other files with date stamps and time stamps in their names, like this:

```
Master.csv.2015-04-06-01-48-32
```

These files are created when a `log` rotate is requested by giving the "`cdr_csv rotate`" command from `fs_cli`, or by sending an HUP signal to the FreeSWITCH process (for example, "`killall -HUP freeswitch`"). This behavior can be changed by setting the following parameter:

```
<param name="rotate-on-hup" value="false"/>
```

Finally, you can specify the path where the `cdr-csv/` directory will be created using the `log-base` parameter. For example, setting `<param name="log-base" value="/var/log"/>` will force all CSV CDR files to be written to the `/var/log/cdr-csv/` directory. (You cannot change the directory's name, which is "cdr-csv". You can change only the location in which it will be created.)

> When changing parameters in `cdr_csv.conf.xml`, be sure to save your changes and then issue the `reload mod_cdr_csv` command at `fs_cli` in order for the changes to take effect.

Other options

There are a few other options in the settings section of `cdr_csv.conf.xml`. The first one is the debug parameter. Setting this to `true` will cause each call to dump all channel variables (like the `info` dialplan application) when the call hangs up. Note that this will dump to both `fs_cli` and the FreeSWITCH log file, so be aware of the disk space.

The other option is called **legs**. This will determine which call leg (or legs) gets a CDR. By default, only the A leg (that is, the calling leg) gets a CDR. You can set this parameter to "b" to log only the B leg (that is, the called leg), or you can set it to "ab" so that you receive a CDR for each leg. Handling A and B legs is discussed later in this chapter.

CDR CSV templates

The `default-template` parameter determines which CDR template is used when creating the CDR record. Notice the `<templates>` section of `cdr_csv.conf.xml`. There are various templates that you can use or edit. You may also create your own templates. By default, we use the example template. Feel free to change or edit the `default-template` parameter to use a different template. The `asterisk` template will output CDRs in the format used by Asterisk PBX. The `sql` template will output records in a particularly useful format, which we will discuss in the *Inserting CDRs into a backend database* recipe.

Templates have another feature that allows custom behavior. When a channel has the `accountcode` variable set to the same name as that of a template, that call's CDR will be formatted in the specified template. You can test this behavior by editing a directory user and setting their accountcode:

1. Open `conf/directory/default/1007.xml` and set this value:

    ```
    <variable name="accountcode" value="sql"/>
    ```

2. Save the file and exit. Issue `reloadxml` at `fs_cli`.

3. Make a test call from 1007 to another phone, answer, and then hang up.

4. You will now have a file named `sql.csv` in your `cdr-csv/` directory. It contains CDR formatted as per the `sql` template defined in `cdr_csv.conf.xml`.

This technique can be used to customize the kinds of data that are stored. For example, you may have a client whose records need to have custom channel variables included in the CDR file, but you may not want every call in your system to include that information. Using an `accountcode` and a CDR CSV template allows you to tailor the behavior as needed.

See also

▶ Refer to the *Inserting CDRs into a backend database* recipe later in this chapter

Using XML CDRs

FreeSWITCH generates a wealth of information for each call that cannot be easily represented in a traditional CSV flat file format. XML gives us all the flexibility to store structured information. In this recipe, we will enable `mod_xml_cdr` and discuss a few of its configuration options.

Getting ready

In the default configuration, `mod_xml_cdr` is compiled but not enabled. Follow these steps to enable it:

1. Open `conf/autoload_configs/modules.conf.xml`.

2. Uncomment this line:

   ```
   <!-- <load module="mod_xml_cdr"/> -->
   ```

3. Save the file and exit.

Now `mod_xml_cdr` will load automatically when FreeSWITCH starts. However, if FreeSWITCH is already running, we need to load it manually. Issue the `load mod_xml_cdr` command at `fs_cli`. The XML CDR data will now be stored in the `log/xml_cdr/` directory.

XML CDRs have many options. To review them, open the `conf/autoload_configs/xml_cdr.conf.xml` file. We will be discussing some of these options later in this recipe.

How to do it...

The easiest way to see a new XML CDR is to use a utility such as `cat` in Linux/Unix or `type` in Windows. Alternatively, you can use a utility such as `less` to `page` through the contents of a file. Both Windows and Linux/Unix support piping the output to `more` to achieve the same effect.

Here are the steps you can follow in a Linux/Unix environment:

1. Change the directory to `/usr/local/freeswitch/log/xml_dr/`.
2. List the directory contents with the `ls` command.
3. Make a test call, perhaps from one phone to another.
4. Hang up the test call and note the new XML CDR, named `a_<uuid>.cdr.xml`.
5. Type `less a_<uuid>.cdr.xml` and press *Enter* to see the content of the XML CDR file.

How it works...

By watching the `log/xml_cdr/` directory, we can observe new CDRs being written to the disk. While not particularly useful in a production system, doing this helps you learn about XML CDRs and the information they contain. Furthermore, it is a simple troubleshooting tool you can use in the future.

What is a UUID?

When dealing with CDRs, and especially XML CDRs, you will be presented with many UUIDs. UUID stands for **Universally Unique Identifier**. It is a string of 32 hexadecimal digits divided into five groups separated by hyphens. An example UUID is 678a195f-8431-4d77-8f10-550f7435f18e. Each call leg receives a UUID in order to keep it distinct from all other call legs.

There's more...

The `mod_xml_cdr` module can do many things, not least posting new XML CDR information to a web server. The web server can then process the XML CDR, whether it means updating a database or performing other billing functions. These are discussed further in the *Using a web server to handle XML CDRs* recipe later in this chapter.

File names and locations

In the `conf/autoload_configs/xml_cdr.conf.xml` file, there are two parameters in the `<settings>` section that affect filenames and locations. The first parameter is called **prefix-a-leg**. When set to true, the A leg XML CDRs will have "a_" prefixed to the filename. This makes it easier to distinguish between A leg and B leg files.

The other parameter is `log-dir`. When set to an absolute path, it will change the location of `/xml_dr/`. Here is an example:

```
<param name="log-dir" value="/var/log"/>
```

This will cause all XML CDRs to be written to the `/var/log/xml_cdr/` directory (you can also set it to a relative path, but that is rarely used).

 When changing parameters in `xml_cdr.conf.xml`, be sure to save your changes and then issue the reload `mod_xml_cdr` command at `fs_cli` in order for the changes to take effect.

Logging the B leg

By default, `mod_xml_cdr` only logs the A leg (that is, the calling leg) of the call. If you wish to log the B leg (that is, the called leg), then set this parameter:

```
<param name="log-b-leg" value="true"/>
```

This will cause the B leg XML CDRs to be written. Note that the B leg CDRs will always be named `<uuid>.cdr.xml`, where `<uuid>` is the actual UUID of the call. There is no option to prefix the filename with "b_" as there is with the A leg.

See also

▸ Refer to the *Using a web server to handle XML CDRs* recipe later in this chapter

Inserting CDRs into a backend database

Frequently, it is necessary to put CDR information into a database such as PostgreSQL or other SQL and NoSQL databases. FreeSWITCH has various modules for writing CDRs directly to many databases, but the preferred architecture is writing CDRs to the disk or posting them to a web server, and then processing them so that they can be inserted into a database. Many engineering reasons lead to this architecture (for example, avoiding dependence on direct, real-time interaction with the database), and most of them relate it to integrity and resilience. This recipe discusses the recommended method of writing SQL-based CSV files and then using those to update a backend database.

Getting ready

Of course, you will need a database in which to store your CDRs. Any SQL-compliant database will work as long as you can use the command line to execute SQL statements. Create a database for your CDRs, and allow any necessary access. This is completely dependent upon the type of database you have. Consult your database documentation for specific instructions.

You will also need a table for the CDRs. The following CREATE TABLE syntax for a PostgreSQL database will work for the existing sql template in cdr_csv.conf.xml:

```
CREATE TABLE cdr (
  caller_id_name character varying(30),
  caller_id_number character varying(30),
  destination_number character varying(30),
  context character varying(20),
  start_stamp timestamp without time zone,
  answer_stamp timestamp without time zone,
  end_stamp timestamp without time zone,
  duration integer,
  billsec integer,
  hangup_cause character varying(50),
  uuid uuid,
  bleg_uuid uuid,
  accountcode character varying(10),
  read_codec character varying(20),
  write_codec character varying(20)
);
```

A similar CREATE TABLE command works for MySQL, as follows:

```
CREATE TABLE cdr (
  caller_id_name varchar(30) DEFAULT NULL,
  caller_id_number varchar(30) DEFAULT NULL,
  destination_number varchar(30) DEFAULT NULL,
  context varchar(20) DEFAULT NULL,
  start_stamp datetime DEFAULT NULL,
  answer_stamp datetime DEFAULT NULL,
  end_stamp datetime DEFAULT NULL,
  duration int(11) DEFAULT NULL,
  billsec int(11) DEFAULT NULL,
  hangup_cause varchar(50) DEFAULT NULL,
  uuid varchar(100) DEFAULT NULL,
  bleg_uuid varchar(100) DEFAULT NULL,
  accountcode varchar(10) DEFAULT NULL,
  domain_name varchar(100) DEFAULT NULL
);
```

All the examples in this recipe will use a database name of `cdr` and a table name of `cdr` as well. The last thing to do is to set the `sql` template as the default CDR template. Follow these steps:

1. Open `conf/autoload_configs/cdr_csv.conf.xml`.

2. Change the `default-template` parameter to `<param name="default-template" value="sql"/>`.

3. Save the file and exit. Issue the reload `mod_cdr_csv` command at `fs_cli`.

4. Issue the `fsctl send_sighup` command at `fs_cli` to rotate the log files.

You are now ready to create and process CDRs.

How to do it...

Follow these steps to get a call record into your new database table:

1. Make a test call from one phone to another, answer, wait a moment, and then hang up (you should now have at least one record in `Master.csv`).

2. Issue the `fsctl send_sighup` command at `fs_cli` (or `cdr-csv rotate`).

3. List the contents of your `log/cdr-csv/` directory, and note the presence of a rotated `Master.csv` file. For example: `Master.csv.2015-04-06-03-37-51`.

4. The rotated `Master.csv` file is the one to use to insert records into your database. You will need to use your specific database's command-line client to insert the records. For PostgreSQL, use a command like this:

    ```
    cat Master.csv.2015-04-06-03-37-51  | tr \" \' |
    psql -U postgres cdr
    ```

5. Confirm the presence of the record in the `cdr` table with a simple SQL query, such as `SELECT * FROM cdr`. Then delete the rotated `Master.csv` file.

How it works...

The `mod_cdr_csv` `sql` template writes CDRs in the format of a single `INSERT SQL` statement per line. A sample record looks like this:

```
INSERT INTO cdr VALUES ("Giovanni Maruzzelli","1002","1005","d
efault","2015-04-06 03:46:00","2015-04-06 03:46:01","2015-04-06
03:46:11","11","10","NORMAL_CLEARING","06d18352-52f3-4d90-836c-
d385a10ea6e3","0a32de98-3318-43aa-8439-c64ddfa9c212", "1002");
```

These INSERT statements can be piped into a database's command-line client. Note the use of tr for conversion of double quotes into single quotes for compatibility with PostgreSQL.

 Your production environment may have specific requirements when it comes to things such as single versus double quotes in PostgreSQL. Using the tr Unix command is one method of handling the issue. You can also modify the template to use single quotes instead of double quotes.

Finally, after confirming that the CDR was successfully inserted into the database, we deleted the rotated file. We can also archive those to another disk volume as a backup.

There's more...

In fact, you can set up a CDR database on a completely separate machine, and use basic tools such as fs_cli to rotate logs and scp or ftp to pull the files to the local database server. An intelligent script can then notify the system administrator of any issues. Also, as long as there is disk space on the FreeSWITCH server, no CDR records will be lost in the event of a failed connection between the CDR server and the FreeSWITCH server. CDRs will continue to be written to the disk on the FreeSWITCH server, and can be collected and processed when connectivity is re-established.

See also

▸ Refer to the *Getting familiar with the "fs_cli" interface* recipe in *Chapter 4, External Control*

Using a web server to handle XML CDRs

One feature of FreeSWITCH's mod_xml_cdr is that it can use HTTP POST actions to send CDR data to a web server, which in turn can process them, and perhaps put them into a database. This mechanism has several advantages:

▸ Modern web servers can handle enormous amounts of traffic

▸ Multiple FreeSWITCH servers can post to a single CDR Server

▸ Multiple web servers can be set up to allow failover and redundancy

It will automatically retry in the event of a failure, on the same web server or on another web server. Eventually, if all the retries on all the web servers fail, it will write to the disk so that the record can be processed later.

The recipe presented here will focus on the steps needed to get a web server set up to process incoming POST requests with XML CDR data.

Getting ready

You will need an operational web server that you control. Most Linux/Unix and Windows systems can get an Apache web server installed. Detailed instructions on configuring a web server are beyond the scope of this book, but such instructions are available in numerous books and on the Internet. This recipe will assume a clean installation of the Apache web server, but the principles apply to other servers as well, such as Lighttpd and Nginx. For this example, we will assume that the Apache server is on the same machine as your FreeSWITCH installation.

How to do it...

Enable `mod_xml_cdr` on your server (refer to the *Using XML CDRs* recipe earlier in this chapter). Next, follow these steps:

1. Open `conf/autoload_configs/xml_cdr.conf.xml` and locate this line:

   ```
   <!-- <param name="url"
           value="http://localhost/cdr_curl/post.php"/> -->
   ```

2. Change it to the following:

   ```
   <param name="url" value="http://localhost/cgi-bin/cdr.pl"/>
   ```

3. Save the file and exit.

4. In your system's `cgi-bin` directory, create a new file named `cdr.pl` (the `cgi-bin` directory is usually at `/usr/lib/cgi-bin`, but it may be different on older systems). Add these lines to the file:

   ```perl
   #!/usr/bin/perl
     use strict;
     use warnings;
     use CGI;
     $|++;
       my $q = CGI->new;
       my $raw_cdr = $q->param('cdr');
       open (FILEOUT,'>','/tmp/cdr.txt');
       print FILEOUT $raw_cdr;
     close(FILEOUT);
   print $q->header();
   ```

5. Save the file and exit.

6. Make the file executable with this command:

   ```
   chmod +x /usr/lib/cgi-bin/cdr.pl
   ```

7. Log in to `fs_cli` and issue the `reloadxml` command. Also, use "reload mod_xml_cdr.".

8. Make a test call. You should see the XML CDR contents in the `/tmp/cdr.txt` file.

How it works...

This is a simple Perl-based CGI script. All it does is pull the `cdr` parameter out of the `POST` data that is submitted by `mod_xml_curl`. Once it has this value (in the `$raw_cdr` variable) it dumps the CDR into a temporary file named `/tmp/cdr.txt`.

While this example is not particularly useful for production, it demonstrates the minimal steps required to get the POSTed CDR data into the system. If you are more comfortable with another scripting language, such as PHP, Python, or Ruby, you may just as easily process the CDRs with that language. Here is a simple version in PHP:

```php
$raw_cdr = $_POST['cdr'];
$writefile = fopen('/tmp/dump.txt',"w");
fwrite($writefile, $raw_cdr);
fclose($writefile);
```

Once you have the data in your program, you can choose how to process it.

There's more...

A common practice with handling XML CDR data with a CGI script (or Fast CGI, or some other appropriate method to handle an HTTP POST request) is to process the data and then put it into a database. This section describes how to insert the CDR into the same database table that we created in the previous recipe, *Inserting CDRs into a backend database*.

Assume that you have a database named `cdr`, with a table also named `cdr`. You can use this modified `cdr.pl` script to insert the records right into the database.

 You will need to use the `cpan` tool to install the **DBI** module and the **DBD** driver for your database. Common ones are `DBD::mysql` and `DBD::PgPP`. This example assumes `DBD::PgPP`, the Postgres "pure Perl" database driver.

The modified `cdr.pl` script is as follows:

```perl
#!/usr/bin/perl
use strict;
use warnings;
use CGI;
```

```
use DBI;
use Data::Dump qw(dump);
$|++;
  my $q = CGI->new;
  my $raw_cdr = $q->param('cdr');
  my @all_fields = qw(caller_id_name caller_id_number
  destination_number context start_stamp answer_stamp end_stamp
  duration billsec hangup_cause uuid bleg_uuid
  accountcode read_codec write_codec);
    my @fields;
    my @values;
    foreach my $field (@all_fields) {
      next unless $raw_cdr =~ m/$field>(.*?)</;
      push @fields, $field;
      push @values, "'" . urldecode($1) . "'";
    }
    my $cdr_line;
    my $query = sprintf(
    "INSERT INTO %s (%s) VALUES (%s);",
    'cdr', join(',', @fields), join(',', @values)
    );
  my $db = DBI->connect('DBI:PgPP:dbname=cdr;host=localhost',
  'postgres', 'postgres');
  $db->do($query);
print $q->header();
sub urldecode {
  my $url = shift;
  $url =~ s/%([a-fA-F0-9]{2,2})/chr(hex($1))/eg;
  return $url;
}
```

This script is a simple example of inserting records into the database. The `@all_fields` array is a list of every field in the `cdr` table. We cycle through this list looking for the corresponding values. If we find one, we use `urldecode` and then add the field name to the `@fields` list, while its value goes into `@values`. From there, we create a query string using the `@fields` and `@values` arrays, and then insert them into the database.

See also

▶ Refer to the *Using XML CDRs and Inserting CDRs into a backend database* recipes covered earlier in this chapter

Using the event socket to handle CDRs

Sometimes, you need alternative ways to get CDR information. FreeSWITCH accommodates those needs with the powerful event socket. This recipe will describe how to receive CDR information about the event socket. You will find more useful information about the event socket interface in the following chapter.

Getting ready

This recipe relies on the event socket interface to FreeSWITCH. There are many different ways of connecting to the event socket. We will use a simple Perl script with the **event socket library** (**ESL**) to demonstrate the principles involved. Any language that supports ESL can use the techniques demonstrated here.

Follow the steps in the *Setting up the event socket library* recipe found in *Chapter 4, External Control*. Specifically, build the Perl module in order to run the example script.

How to do it...

Enter this script (or download it from the Packt Publishing website at http://www.packtpub.com):

```perl
# Connect to event socket, listen for CHANNEL_HANGUP_COMPLETE events
# Uses event data to create custom CDRs
use strict;
use warnings;
use lib '/usr/src/freeswitch.git/libs/esl/perl';
use ESL;
my $host = "localhost";
my $port = "8021";
my $pass = "ClueCon";
my $con  = new ESL::ESLconnection($host, $port, $pass);
if ( ! $con ) {
        die "Unable to establish connection to FreeSWITCH.\n";
}
## Listen for events, filter in only CHANNEL_HANGUP_COMPLETE
$con->events('plain','all');
$con->filter('Event-Name','CHANNEL_HANGUP_COMPLETE');
print "Connected to FreeSWITCH $host:$port and waiting for
events...\n\n";
while (1) {
        my $e = $con->recvEvent();
        my @raw_data = split "\n",$e->serialize();
```

```perl
my %cdr;
foreach my $item ( @raw_data ) {
        #print "$item\n";
        my ($header, $value) = split ': ', $item;
        $header =~ s/^variable_//;
        $cdr{$header} = $value;
}
# %cdr contains a complete list of channel variables
print "New CDR: ";
print $cdr{uuid} . ', ' . $cdr{direction} . ', ';
print $cdr{answer_epoch} . ', ' . $cdr{end_epoch} . ', ';
print $cdr{hangup_cause} . "\n";
}
```

Run this script and make a test call. An abbreviated CDR will be printed to the screen. Press *Ctrl + C* to exit the script.

How it works...

The basic principles involved are as follows:

- Establish an ESL connection to FreeSWITCH
- Subscribe to the CHANNEL_HANGUP_COMPLETE events using a filter
- Process each event as an individual CDR

If you are more familiar with PHP, Python, or Ruby, you should be able to translate these concepts from our demonstration script.

There's more...

Here are a few tips to help you make the most of using the event socket for CDRs.

ESL considerations

Keep in mind that the script will need to be able to find the ESL library. Note this line:

```perl
use lib '/usr/src/freeswitch.git/libs/esl/perl';
```

This tells Perl to look in the specified directory when using additional modules. Without it, the use of the ESL directive will fail (alternatively, you can install the requisite ESL files in your system's site Perl directory).

Another important point is that this method will receive two events for a normal bridged call. The A leg and the B leg each generate a CHANNEL_HANGUP_COMPLETE event. The value in $cdr{direction} will be *inbound* for the A leg and *outbound* for the B leg.

Finally, keep in mind that this line is blocking:

```
my $e = $con->recvEvent();
```

It will block the entire script until a new event arrives. See the *Filtering events* recipe in *Chapter 4, External Control* to see an example of the `recvEventTimed()` method that does not block.

Receiving XML CDRs

It is possible to receive the CDRs over the event socket in XML format. This is controlled on a per-call basis using the `hangup_complete_with_xml` channel variable. Set this variable to `true` in your dialplan as follows:

```
<action application="set" data="
hangup_complete_with_xml=true"/>
```

See also

▸ Refer to the *Using XML CDRs* recipe in this chapter for more information on XML-based CDRs

▸ Refer to the *Setting up the event socket library* and *Filtering events* recipes in *Chapter 4, External Control*

Directly inputting CDRs into various databases in real time

Although various engineering reasons militate against it, many in the community felt the need for FreeSWITCH to directly write CDRs to database tables.

How to do it...

Various modules give you flexibility of SQL and NoSQL data storage:

▸ `mod_cdr_mongodb`: This saves detailed CDR data in a MongoDB database, in a format similar to `mod_json_cdr`.

▸ `mod_odbc_cdr`: This saves any channel variable from the call to an ODBC database of your choice.

▸ `mod_cdr_pg_csv` : This logs call detail records (CDRs) directly to a PostgreSQL database, using the schema defined in the `config` file.

- ▶ `mod_cdr_sqlite`: This saves directly in a sqlite DB with the variables you specify in a template.

- ▶ `mod_json_cdr`: This saves in the file or POSTs a JSON representation of the channel variable and call flow. It can post directly to CouchDB.

- ▶ `mod_radius_cdr`: This is the RADIUS CDR module.

There's more...

As always, you can find plenty of information on the FreeSWITCH documentation site, starting off from `http://freeswitch.org/confluence/display/FREESWITCH/CDR`.

4
External Control

In this chapter, we will cover the following recipes:

- ▶ Getting familiar with the `fs_cli` interface
- ▶ Setting up the event socket library
- ▶ Establishing an inbound event socket connection
- ▶ Establishing an outbound event socket connection
- ▶ Using `fs_ivrd` to manage outbound connections
- ▶ Filtering events
- ▶ Launching a call with an inbound event socket connection
- ▶ Using the ESL connection object for call control
- ▶ Using the built-in web interface

Introduction

One of the most powerful features of FreeSWITCH is the ability to connect to it and control it from an external resource. This is made possible by the powerful FreeSWITCH event system and its connection to the outside world—the event socket. The event socket interface is a simple TCP-based connection that programmers can use to connect to the inner workings of a FreeSWITCH server. Furthermore, the developers of FreeSWITCH have also created the **Event Socket Library** (**ESL**), which is an abstraction layer meant to make programming with the event socket a lot simpler. The following languages are supported by ESL:

- ▶ C/C++
- ▶ Lua
- ▶ Perl

- PHP
- Python
- Ruby
- Java
- Managed
- TCL

Keep in mind that ESL is only an abstraction library. You can connect to the event socket with any socket-capable application, including **telnet**!

The tips in this chapter will focus mostly on using the event socket for some common use cases. The last tip, however, will introduce a particularly interesting way to connect to FreeSWITCH externally without using the event socket, that is, using the built-in web server that is enabled when you install mod_xml_rpc. Regardless of how you wish to control FreeSWITCH, it is highly recommended that you read the first recipe in this chapter, *Getting familiar with the fs_cli interface*, as this will serve you well in all aspects of working with FreeSWITCH.

Getting familiar with the fs_cli interface

The preferred method of connecting to the FreeSWITCH console is by using the fs_cli program, where fs_cli stands for **FreeSWITCH Command-line Interface**. This program comes with FreeSWITCH as part of the default installation, and works with Linux/Unix, Mac OS X, and Windows. What is less known about fs_cli is that it is an excellent example of an ESL program. Beyond that, anything that you can do with fs_cli can also be done with ESL and the event socket.

Keep in mind that when you are logged in to fs_cli, you can do anything that you can do on the FreeSWITCH console, including shutting down the system and disconnecting any calls. So, exercise appropriate caution when using fs_cli.

Naturally, the first step in mastering external control of FreeSWITCH is to become familiar with fs_cli. Indeed, it is one of the most important tools for interacting with your FreeSWITCH server.

If you're familiar with C programming, then you might appreciate the source code for fs_cli. It is found in libs/esl/fs_cli.c under the FreeSWITCH source directory.

Getting ready

The only prerequisites for running `fs_cli` are access to your system's command line and a running FreeSWITCH server with `mod_event_socket` enabled (in the case of a default installation, `mod_event_socket` is always enabled). However, you may find it convenient to allow `fs_cli` to be launched from any directory on your system. In a Linux/Unix environment, you can add a symbolic link, such as this:

```
ln -s /usr/local/freeswitch/bin/fs_cli /usr/local/bin/fs_cli
```

Windows users can add the FreeSWITCH binary directory to their system's `PATH` variable.

How to do it...

Follow these steps:

1. Launch `fs_cli` by typing `fs_cli` (or in Windows, `fs_cli.exe`) and pressing *Enter*. A simple welcome screen will appear, as follows:

At this point, you are in `fs_cli` and can issue commands.

2. Try a simple command; type `/help` and press *Enter*. You will see a number of commands that you can enter. All commands that begin with a forward slash (`/`) are specific to the `fs_cli` program (that is, do not go to the FreeSWITCH server).

3. You can also issue FreeSWITCH API commands. Type `show api` and press *Enter*. You will see quite a long list of available FreeSWITCH API commands.

4. Finally, type `status` and press *Enter* to see a brief status report about your FreeSWITCH server.

How it works...

The `fs_cli` emulates the behavior of the FreeSWITCH console, which is available when FreeSWITCH is run in the foreground, that is, without the `-nc` flag ("nc" stands for "no console"). However, technically speaking, `fs_cli` is merely an event socket program. Everything sent and received with `fs_cli` is done over the FreeSWITCH event socket. Therefore, just about everything that you can do from `fs_cli` can also be done with an event-socket-based program. Keep in mind that the "slash" commands are specific to `fs_cli` (that is, don't communicate with FreeSWITCH) and don't necessarily have an event socket equivalent, such as `/help` and `/exit`.

> There are numerous ways to exit the `fs_cli` program. There are three equivalent "slash" commands, namely `/exit`, `/quit`, and `/bye`. You can also type three periods (`. . .`) and press *Enter*. On some systems, you can press *Ctrl + D*.

There's more...

Now that you are familiar with the general usage of `fs_cli`, it will be good to learn about some of the more useful commands.

Important commands for listing information

FreeSWITCH administration frequently means getting information from the server. Here is a list with brief descriptions of some commands that you will, no doubt, want to use. Feel free to try any of these on your system. They won't "break" anything! They will simply give you information.

Command	Description
`sofia status`	Display general SIP information
`sofia status profile internal`	Display SIP information about the "internal" profile
`help`	List the available commands (equivalent to `show api`)

Command	Description
show channels	List individual call legs
show calls	List bridged calls
/log 6	Set the log level to INFO (this prevents the numerous "debug" messages from being displayed)
/log 7	Set the log level to DEBUG (all the "debug" messages are displayed)

Useful command-line options

The fs_cli program has a number of command-line options. You can view them all by executing fs_cli -h (or fs_cli.exe -h in Windows). The following are descriptions of some of the more useful options:

Option	Description
-x	Execute a command and then exit.
-r	Retry the connection until it succeeds (This is useful if you have just restarted FreeSWITCH. It will retry until FS completes startup and answers you).
-R	This will automatically reconnect if unwillingly disconnected.
-u	Include the uuid in the output (for example, which call uuid generated the output line).
-s	Include the short uuid in the output (such as -u, uuid in shortened format).
-H	Specify the FreeSWITCH server hostname or IP address to connect to.
-P	Specify the FreeSWITCH server port to connect to.

The -x option is particularly useful for doing things from the command line and quick scripts. For example, try this command from your system's Command Prompt:

```
fs_cli -x "show channels"
```

You will receive the output from the show channels command and then come back to the command shell.

 See the online documentation for fs_cli at http://freeswitch.org/confluence/display/FREESWITCH/Command-Line+Interface+fs_cli. It includes descriptions of all the fs_cli commands, as well as the handy .fs_cli_conf configuration file.

Viewing events

Event-based programming can be a daunting challenge at first. As a brief introduction, it is good just to look at the events that come over the event socket. The `fs_cli` can do this very easily. At `fs_cli`, enter these two commands:

```
/log 0
```

```
/event plain all
```

Watch your screen for a few seconds, and you will eventually see some events come in. Whenever a call is handled on the system, there will be numerous events. There are events for changes in the call state, as well as new calls being set up and existing calls being torn down. If anything happens, then there will be at least periodic HEARTBEAT and RE_SCHEDULE events. Issue the `/noevents` command to stop seeing the events come through.

> Remember that you can always use `fs_cli` as an additional monitor for events when you're interacting via ESL.

The rest of this chapter contains a great deal of information about event socket programming.

See also

▸   Refer to the *Filtering events* recipe in this chapter

Setting up the event socket library

Most event socket programming is not usually done in C, but rather in one of the more common scripting languages, such as Perl, PHP, Python, Ruby, and so on. The **Event Socket Library** (**ESL**) is available as a tool for abstracting the nitty-gritty of socket interacting.

Getting ready

The most difficult part about using ESL with a scripting language is making sure that the necessary development libraries have been installed. This process varies among operating systems and languages. The instructions presented here are for Debian or Red Hat Linux variants. If your operating system is not among these, then it is recommended that you check with the website for your language and look for instructions on how to install the development libraries.

Debian/Ubuntu

Debian variants (such as Ubuntu) generally use the `apt` package manager. The development libraries can be installed with these commands:

```
apt-get install libxml2-dev libpcre3-dev libcurl4-openssl-dev libgmp3-
dev libaspell-dev python-dev php5-dev libonig-dev libqdbm-dev libedit-dev
libperl-dev
```

Red Hat/CentOS

Red Hat Linux variants (RHEL, CentOS, and Fedora) generally use the yum package manager. The development libraries can be installed with these commands:

```
yum install libxml2-devel pcre-devel bzip2-devel curl-devel gmp-devel
aspell-devel libtermcap-devel gdbm-devel db4-devel libedit-devel php-
devel ruby-devel lua-devel perl-devel python-devel
```

How to do it...

All the libraries and tools must be installed before the initial configuration of the FreeSWITCH sources (that is, before compiling FS). If you add them later, you *must* ignore the error if `config.cache` does not exist:

```
cd /usr/src/freeswitch
```

```
rm config.cache
```

```
./config.status --recheck && ./config.status
```

Once you have the necessary library files installed for your language of choice, you are ready to do the actual build of ESL. Open a terminal and change the directory to your FreeSWITCH source directory. From there, execute these commands:

```
cd libs/esl
```

```
make
```

This will confirm that your system's ESL libraries are ready to be used. From here, you can install the library for your language of choice. In our example, we'll use Perl. Execute this command:

```
make perlmod-install
```

You can also use other languages using one of the following commands:

- `make phpmod-install`
- `make pymod-install`
- `make rubymod-install`

Once the installation is complete, you can start using ESL in your scripts.

How it works...

The make install commands will compile ESL for each language, and then install the necessary files on the language's include path. If you have a nonstandard installation, then you may have to install these files yourself. Once the files are installed, you can use one of the test scripts for your language. For example, you can change the directory to the perl/ subdirectory, and run the single_command.pl script for testing:

cd perl

perl single_command.pl status

The other languages have sample scripts as well. Run a sample script to confirm that your ESL is working. From here, you can move on to perform other ESL-related tasks.

> If you get an error such as **Can't call method "getBody" on an undefined value**, then it means that FreeSWITCH is most likely not running. Make sure that FreeSWITCH is running and also that you can connect to it using fs_cli.

Establishing an inbound event socket connection

An inbound event socket connection means that an external script or program is connecting to a FreeSWITCH server. The connection is inbound from the server's point of view. In fact, every time you run the fs_cli utility, you are making an inbound event socket connection.

Getting ready

Be sure that you have installed ESL for your preferred programming language (see the previous recipe, *Setting up the event socket library*). From there on, you will just need a text editor, command-line access, and a phone registered to your system. The examples presented here are in Perl. However, the accompanying code samples have corresponding examples in Python as well.

How to do it...

The following code is a simple inbound connection that sends the status command to FreeSWITCH. Add the code as follows:

1. Open scripts/ib_api.pl in a text editor and add these lines:

   ```
   #!/usr/bin/perl
   ```

```perl
use strict;
use warnings;
require ESL;

my $host = "127.0.0.1";
my $port = "8021";
my $pass = "ClueCon";
my $con  = new ESL::ESLconnection($host, $port, $pass);
if (! $con) {
    die "Unable to establish connection to $host:$port\n";
}
my $cmd  = "status";
my $args = "";
my $e    = $con->api($cmd, $args);

if ( $e ) {
    print "Result of $cmd $args command:\n\n";
    print $e->getBody();
} else {
    die "No response to $cmd command.\n";
}
```

2. Save the file and exit.

3. Linux/Unix users can make the script executable with this command:

   chmod +x ib_api.pl

4. Run the script and you will see the output of the status command:

   Linux/Unix: ./ib_api.pl

   Windows: perl.exe ib_api.pl

```
root@vz124:/usr/local/freeswitch/scripts# ./ib_api.pl
Result of status   command:

UP 0 years, 1 day, 3 hours, 53 minutes, 33 seconds, 997 milliseconds, 264 micros
econds
FreeSWITCH (Version 1.4.18 git 4eed221 2015-03-12 18:55:23Z 64bit) is ready
0 session(s) since startup
0 session(s) - peak 0, last 5min 0
0 session(s) per Sec out of max 30, peak 0, last 5min 0
1000 session(s) max
min idle cpu 0.00/99.53
Current Stack Size/Max 240K/8192K
root@vz124:/usr/local/freeswitch/scripts# _
```

How it works...

The script basically does these four things:

- Uses (that is, `requires`) the ESL library
- Connects to FreeSWITCH with the `ESL::ESLconnection` object
- Issues the `status` command with the connection object's `api()` method
- Prints the results with the event object's `getBody()` method

Change the `$cmd` and `$args` values to issue a different command. For example, to see the results of `sofia status profile internal` you have to set the variables as follows:

```
my $cmd  = "sofia";
my $args = "status profile internal";
```

Note that we also perform some very basic error checking. First, we confirm that we are getting a valid `ESL::ESLconnection` object. Secondly, we make sure that we receive an event object as a result of the `$con->api` call.

There's more...

The ESL event object has a number of methods. One of the most important is the `getBody()` method. However, not all events actually have a body—they simply have a list of headers. To see what the event headers look like, use the `serialize()` method, as follows:

```
print $e->serialize();
```

This will print a list of headers and their corresponding values. Try it! You can also get an individual header value with the `getHeader()` method:

```
print $e->getHeader('Event-Name');
```

Keep in mind that we are using the `api()` method, which blocks (that is, it waits for a response). This keeps things simple, but there are times when blocking is not desired. The `ESL::ESLconnection` object also has a `bgapi()` method for executing API commands in a non-blocking manner. The `bgapi()` method is discussed further in the *Launching a call with an inbound event socket connection* recipe later in this chapter.

See also

- Refer to the *Setting up the event socket library* and *Launching a call with an inbound event socket connection* recipes in this chapter

Establishing an outbound event socket connection

An **outbound** event socket connection lets you control a call leg from a program that sits and waits for a TCP connection on a specific port. The dialplan socket application sends control of the call to the process listening on the specified TCP port. This recipe will guide you through the steps necessary to get a simple call control script up and running. You may find it easier to understand the information presented here if you are at least somewhat familiar with the concept of TCP sockets.

Getting ready

You will need a text editor and a telephone connected to FreeSWITCH, as well as access to the `fs_cli` for your system. You will also need to have ESL compiled and working for your scripting language of choice (see *Setting up the event socket library* earlier in this chapter). The language used in this is example is Perl. However, the principles apply to all ESL-enabled languages. When we are through, we will have a simple script that will listen for a socket connection from FreeSWITCH, answer the call, play a file, wait for a DTMF digit, and then exit.

How to do it...

Start by creating an extension to dial:

1. Open `conf/dialplan/default/01_Custom.xml` in a text editor, and add this simple extension:

   ```xml
   <extension name="outbound event socket">
     <condition field="destination_number" data="^(5004)$">
       <action application="socket" data="127.0.0.1:8040 async"/>
     </condition>
   </extension>
   ```

2. Save the file and exit. Issue the `reloadxml` command at `fs_cli`.

   Now create the script.

3. Create the `scripts/outbound_socket.pl` file in a text editor, and add these lines:

   ```perl
   #!/usr/bin/perl
   require ESL;
   use IO::Socket::INET;

   my $ip = "127.0.0.1";
   ```

```perl
    my $sock = new IO::Socket::INET ( LocalHost => $ip,
                                      LocalPort => '8040',
                                      Proto => 'tcp',
                                      Listen => 1,
                                      Reuse => 1 );
die "Could not create socket: $!\n" unless $sock;

for(;;) {
    my $new_sock = $sock->accept();
    my $pid = fork();
    if ($pid) {
        print "New child pid $pid created...\n";
        close($new_sock);
        next;
    }

    my $fd = fileno($new_sock);
    my $con = new ESL::ESLconnection($fd);
    my $info = $con->getInfo();
    my $uuid = $info->getHeader("unique-id");

    printf "Connected call %s, from %s\n", $uuid,
            $info->getHeader("caller-caller-id-number");

    $con->sendRecv("myevents $uuid");
    $con->setEventLock("1");
    $con->execute("answer");
    $con->execute("start_dtmf");
    $con->execute("playback",
                    "ivr/ivr-welcome_to_freeswitch.wav");
    $con->execute("sleep","500");
    $con->execute("playback",
                    "ivr/ivr-finished_pound_hash_key.wav");

    while($con->connected()) {
        my $e = $con->recvEvent();
        if ($e) {
            my $name = $e->getHeader("event-name");
            print "EVENT [$name]\n";
            if ($name eq "DTMF") {
                my $digit = $e->getHeader("dtmf-digit");
                my $duration = $e->getHeader("dtmf-duration");
                print "DTMF digit $digit ($duration)\n";
```

```
                    $con->execute("hangup");
                }
            }
        }
        print "BYE\n";
        close($new_sock);
    }
```

4. Save the file and exit.

5. Linux/Unix users make the script executable with this command:

    chmod +x outbound_socket.pl

6. Launch the script, as follows:

    Linux/Unix: `./outbound_socket.pl`

    Windows: `perl.exe outbound_socket.pl`

7. The script is now waiting for a connection. Dial `5004` from your phone and watch the script's output to see what it is doing.

outbound_socket.pl controlling an incoming call

How it works...

The script opens a `socket listener` on the localhost IP address of 127.0.0.1 and TCP port 8040. When you call `5004`, it executes the `socket` application, which literally sends the control of the call over to port 8040. The `socket` application has no idea what is listening on that port or even whether there is anything listening (try dialing `5004` without the script running).

Once the socket connection is opened, the Perl script "forks" a "child process" and continues to listen for further connections (if we didn't do this, then the script would exit after the first call it handled, and we would need to restart it after each call). If the fork is successful, then the new child process executes the code, starting with this line:

```
my $fd = fileno($new_sock);
```

Most of these lines are fairly obvious, but a few of them warrant some explanation. Let's start with these lines:

```
my $fd  = fileno($new_sock);
my $con = new ESL::ESLconnection($fd);
```

The `$fd` variable is a file descriptor for the socket connection that is opened. It is passed to the new method of the `ESL::ESLconnection` object class to ensure that the `$con` object communicates with the correct TCP stream from FreeSWITCH. Once we have the connection object (`$con`), we can get some information from it, with these lines:

```
my $info = $con->getInfo();
my $uuid = $info->getHeader("unique-id");
```

The `$info` object is a representation of the initial burst of information that FreeSWITCH sends to the script when the socket connection is first established. The `$uuid` variable is populated with the call leg's UUID, which is found in the `unique-id` header of the `$info` object.

This line is important for outbound socket connections:

```
$con->sendRecv("myevents $uuid");
```

By default, the socket will not get any event. The `myevents` command is a special event socket directive that tells FreeSWITCH that this particular socket session will receive events only for this particular call leg. In effect, it filters out all FreeSWITCH events that do not pertain to that call leg (ask for "event plain ALL" instead to subscribe to ALL events). The `sendRecv` method sends an event socket command and waits for a response. Note that `sendRecv` is very different from the `execute` method. The `execute` method executes a dialplan application, whereas the `sendRecv` command sends an event socket command.

Then we find this:

```
$con->setEventLock("1");
```

The `setEventLock` is required here to respect the order of execution, which—because we're using an async socket—would be random. In this way, we can have all the advantages of an async execution and maintain the sequence of instructions.

We use the `execute` method to play a few sound files, and then we enter this important `while` loop:

```
while($con->connected()) {
    my $e = $con->recvEvent();
    if ($e) {
        ...
        }
    }
}
```

This control structure checks two things: the status of the connection, and whether or not an event has been received. If the caller hangs up, then `$con->connected()` will evaluate to `false` and the script will exit. Also, if the user presses a touch tone, then the script will receive an event. The script is receiving other events as well, but we ignore anything that is not a DTMF key press.

Finally, if we receive an event, then the `$e` object is populated. Now we can check whether it is a DTMF event:

```
my $name = $e->getHeader("event-name");
print "EVENT [$name]\n";
  if ($name eq "DTMF") {
```

For each event received, we print the name of the event. However, we act only upon receiving a `DTMF` event. We display some information about the DTMF that was received, and then hang up the call.

There's more...

When the event socket connection is first made, FreeSWITCH sends an initial burst of information to the script. To see what this looks like, add the following line right after the `printf` line:

print $info->serialize();

Make the call to `5004` again, while watching the script's output. You will see that there is a tremendous amount of information that FreeSWITCH sends when the call is first established. Use the `getHeader()` method to retrieve a specific value from the `$info` object, as we did with `unique-id`.

▶ For an alternative way of handling multiple connections, see the *Using fs_ivrd to manage outbound connections* recipe in this chapter, which discusses a special utility to make the job easier

▶ Also refer to the *Setting up the event socket library* and *Using the ESL connection object for call control* recipes in this chapter

Using fs_ivrd to manage outbound connections

FreeSWITCH supplies a tool that offers a simplified means of creating interactive scripts. Unlike the socket application presented in the *Using the ESL connection object for call control* recipe in this chapter, using `fs_ivrd` relieves the programmer from having to maintain socket connections and handle child processes. The `fs_ivrd` tool provides a simple interface using the `STDIN` and `STDOUT` file handles. The example Perl script presented here uses the `ESL::IVR` Perl module supplied with ESL.

Getting ready

This example requires that the ESL Perl module be properly compiled and installed. See the *Setting up the event socket library* recipe earlier in this chapter. Also, it is helpful to have at least two terminal windows open so that you can view the script as well as `fs_cli`. Note that `fs_ivrd` is not supported in Windows environments.

How to do it...

First, add a new extension to your dialplan by following these steps:

1. Edit or create a new file in `conf/dialplan/default/` named `01_event_socket.xml`.

2. Add this extension to the new file:

```
<extension name="fs_ivrd Example">
  <condition field="destination_number" expression="^(9950)$">
    <action application="log"
            data="INFO Starting fs_ivrd example..."/>
    <action application="set" data=
      "ivr_path=/usr/local/freeswitch/scripts/ivrd-example.pl"
      />
    <action application="socket" data="127.0.0.1:9090 full"/>
  </condition>
</extension>
```

3. Save the file, exit, and then issue `reloadxml` or press *F6* at the `fs_cli` prompt.

   This extension will call your `fs_ivrd` script when the user dials `9950`. Create the following script, or download it from the Packt Publishing website:

4. Create a new file in `scripts/` called `ivrd-example.pl`.

5. Add the following lines to it:

```perl
#!/usr/bin/perl
use strict;
use warnings;
use ESL::IVR;

$| = 1;          # Turn off buffering
select STDERR; # Use this stream for console output
print "Starting ivrd-example.pl...\n\n";

my $con = new ESL::IVR;
my $uuid = $con->{_uuid};
my $dest = $con->getVar('destination_number');

$con->execute('answer');
$con->execute('sleep','500');
$con->playback('ivr/ivr-welcome_to_freeswitch.wav');
my $digits = "1";
my $prompt = 'file_string://voicemail/vm-to_exit.wav';
$prompt .= '!voicemail/vm-press.wav!digits/9.wav';
my $badinput = 'ivr/ivr-that_was_an_invalid_entry.wav';

while( $con->{_esl}->connected() ) {
  while ( $con->{_esl}->connected() && $digits != "9" ) {
    $con->playAndGetDigits(
         "1 1 3 5000 # $prompt $badinput mydigits \\d+");
    $digits = $con->getVar('mydigits');
    print "Received digit $digits\n";
    $con->playback("ivr/ivr-you_entered.wav");
    $con->execute("say","en number pronounced $digits");
    $con->execute("sleep","1000");
    if ( $digits == "9" ) {
      $con->playback('voicemail/vm-goodbye.wav');
    }
  }
}
  $con->execute("hangup");
}
```

6. Save the file and exit.

7. Make the file executable with this command:

   `chmod +x ivrd-example.pl`

8. Finally, we need to launch the `fs_ivrd` daemon with this command:

   `/usr/local/freeswitch/bin/fs_ivrd -h 127.0.0.1 -p 9090`

9. Test the script by dialing `9950` and following the prompts.

How it works...

The `fs_ivrd` daemon runs constantly. In fact, you can run it in the background using any `bg` command that is appropriate for your platform. When it receives a socket connection from FreeSWITCH, it launches whatever script is specified in the `ivr_path` channel variable, and handles all inter-process communications. The `ivrd-example.pl` script simply establishes an ESL connection using the `ESL::IVR` module. The resulting `$con` object is a superset of the standard ESL connection object.

Once the connection is made, the actual call control is quite simple: we answer the call, pause, and then greet the caller. We then enter an outer `while` loop that checks whether or not the caller has hung up. The inner `while` loop checks for two conditions:

▶ Whether or not the caller has hung up

▶ Whether the caller has dialed `9`

If either case is true, then the script exits. Otherwise, we simply ask the caller to press a digit, read it back, and loop around again.

Building custom, interactive call control scripts with `ESL::IVR` is all but simple. Just use the `ivrd-example.pl` script as a template. Note that your script can also use any other Perl modules available on your system, such as the DBI module, for database access.

See also

▶ The *Establishing an outbound event socket connection* and *Using the ESL connection object for call control* recipes in this chapter

Filtering events

Events are the lifeblood of the FreeSWITCH eventing system. FreeSWITCH throws events for virtually everything that happens. This can overwhelm a program (and indeed, the programmer) with a flood of information. The solution is to use the FreeSWITCH event filter feature.

Getting ready

Learning about filters is very simple. Initially, we will just use an `fs_cli` connected to a FreeSWITCH server. Later, we will look at some simple programming examples using ESL. You will need a phone connected to your FreeSWITCH server and two terminal windows open so that you can look at your program in one session and `fs_cli` in another.

How to do it...

Consider a simple example. Here, we will compare the event socket output before and after using a filter:

1. Launch `fs_cli` and connect to a running FreeSWITCH server. Issue these two commands at `fs_cli`:

    ```
    /log 0
    /event plain all
    ```

2. Wait a few seconds. No doubt, you'll see some events, and possibly a lot of events.

3. From your phone, dial *98 and wait for the system to answer. Then hang up. You should see many events.

4. Let's filter out everything except the channel hang up events. Issue this command:

    ```
    /filter Event-Name CHANNEL_HANGUP_COMPLETE
    ```

5. Repeat the call to *98 and then hang up. You should see only a single event.

How it works...

FreeSWITCH uses a `filter in` system (as opposed to a `filter out` system) to filter events. If no filters have been set, then the event socket shows all events. The command we issued means, in effect, "Show all CHANNEL_HANGUP_COMPLETE events." You may set additional filters, like this for example:

```
/filter Event-Name CHANNEL_HANGUP_COMPLETE
```

```
/filter Event-Name CHANNEL_EXECUTE
```

These commands add two filters. In effect, they mean, "Show all CHANNEL_HANGUP_COMPLETE events and all CHANNEL_EXECUTE events." There is no limit to the number of filters you may set on an event socket connection.

The `fs_cli` is useful for looking at simple events and performing some basic debugging, but in practice you will probably need to apply filters from within a program.

Consider this functional Perl script:

```perl
use ESL;
my $con = new ESL::ESLconnection("localhost", "8021", "ClueCon");
if ( !$con ) {
   die "Unable to connect to FreeSWITCH server; $!\n";
}
$con->events('plain','all');
while (1) {
my $e = $con->recvEventTimed(10);
next unless $e;
print $e->serialize();
}
```

Although not particularly useful, this Perl script demonstrates how to connect to the FreeSWITCH event socket using ESL and listening for events. When it receives an event, it will print it on the console. The $con variable is the ESL connection object, and $e is an event object. Run this script on your system, and you will see that it dumps every event. Let's add a filter and a few strategic print statements. Modify the script as follows:

```perl
$con->events('plain','all');
$con->filter('Event-Name','CHANNEL_STATE');
while (1) {
   my $e = $con->recvEventTimed(10);
   next unless $e;
   my $chan_state = $e->getHeader('Channel-State');
   my $chan_call_state = $e->getHeader('Channel-Call-State');
   my $chan_leg = $e->getHeader('Call-Direction') eq 'inbound' ? 'A' :
   'B';
   my $chan_name = $e->getHeader('Channel-Name');
   print "($chan_leg Leg) $chan_state / $chan_call_state
   [$chan_name]\n";
}
```

First, note that we add a filter on CHANNEL_STATE events. This will let us receive events only when there is a state change on a channel, for example, when a channel goes from "ringing" to "answered." We also create several Perl variables, as follows:

Variable	Purpose
$chan_state	Channel state (NEW, INIT, and ROUTING)
$chan_call_state	Call state (RINGING, ACTIVE, HANGUP, and DOWN)
$chan_leg	Call leg (A leg or B leg)
$chan_name	Channel name

Run this script on your system, and then make a call from one phone to another. Watch the output while the target phone is ringing, then when the target phone is answered, and finally when one of the phones hangs up. Observing this process will help you grasp the types of events that FreeSWITCH throws as calls traverse the system.

See also

▶ The *Setting up the event socket library* recipe earlier in this chapter

Launching a call with an inbound event socket connection

Using an inbound event socket connection to launch a call is a common requirement for some applications, such as outbound IVRs. In a case such as this, it is advantageous to handle the generation of calls in a nonblocking manner using the ESL connection object's `bgapi()` method. This recipe discusses how to use the `bgapi()` method with the corresponding "Background-Job UUID."

Getting ready

Be sure that you have configured ESL for your system and that you have followed the steps in the *Establishing an inbound event socket connection* recipe earlier in this chapter. The examples here are written in Perl, but the principles apply to any ESL-enabled language. Of course, you will need a text editor and a SIP phone registered to your FreeSWITCH server in order to test this example.

How to do it...

Start by creating the new script:

1. Create the `scripts/ib_bgapi.pl` file in a text editor, and add these lines:

```perl
#!/usr/bin/perl
use strict;
use warnings;
require ESL;

my $host = "127.0.0.1";
my $port = "8021";
my $pass = "ClueCon";
my $con  = new ESL::ESLconnection($host, $port, $pass);
```

```
        if (! $con) { die "Unable to establish connection to
        $host:$port\n"; }
        $con->events("plain","all");

        my $target = shift;
        my $uuid = $con->api("create_uuid")->getBody();
        my $res =
          $con->bgapi("originate","{origination_uuid=$uuid}$target 9664");
        my $job_uuid = $res->getHeader("Job-UUID");
        print "Call to $target has Job-UUID of $job_uuid and call uuid of
        $uuid \n";

        my $stay_connected = 1;
        while ( $stay_connected ) {
          my $e = $con->recvEventTimed(30);
          if ( $e ) {
            my $ev_name = $e->getHeader("Event-Name");
            if ( $ev_name eq 'BACKGROUND_JOB' ) {
              my $call_result = $e->getBody();
              print "Result of call to $target was $call_result\n\n";
            } elsif ( $ev_name eq 'DTMF' ) {
              my $digit = $e->getHeader("DTMF-Digit");
              print "Received DTMF digit: $digit\n";
              if ( $digit =~ m/\D/ ) {
                print "Exiting...\n";
                $stay_connected = 0;
              }
            } else {
              # Some other event
            }
          } else {
            # do other things while waiting for events...
          }
        }
        $con->api("uuid_kill",$uuid);
```

2. Save the file and exit.

3. Linux/Unix users make the script executable with this command:

   chmod +x ib_bgapi.pl

4. Launch the script, as follows:

   Linux/Unix: `./ib_bgapi.pl user/1002`

   Windows: `perl.exe ib_bgapi.pl user/1002`

Be sure to replace `1002` with the extension number for your phone. Your phone should ring; when you answer, you will hear some music. Watch the console as you answer the call and press the DTMF digits. Press * or # to exit the script.

```
root@vz124:/usr/local/freeswitch/scripts# ./ib_bgapi.pl user/1002
Call to user/1002 has Job-UUID of a4f7426e-50c3-4448-a0e1-13a60ce3d631 and call
uuid of b14b8fe5-a8fb-4bc0-8d3a-b7d6f794c397
Result of call to user/1002 was +OK b14b8fe5-a8fb-4bc0-8d3a-b7d6f794c397

Received DTMF digit: 9
Received DTMF digit: 1
Received DTMF digit: 9
Received DTMF digit: *
Exiting...
root@vz124:/usr/local/freeswitch/scripts#
```

How it works...

This script takes a dial string as an argument on the command line, and then makes a `bgapi` (background API) origination attempt to that dial string. Whenever `bgapi` is called, there will always be a "Job-UUID" response. The `bgapi` command is discussed a little later. We use the `uuid_create` method of the ESL connection object to create a UUID that we can assign to our outbound call leg. Normally, FreeSWITCH will assign a UUID value to each call leg. However, by preselecting the UUID value, we save ourselves some extra (unnecessary) parsing of events to try to recover the UUID.

At this point, we generate the outbound call, print some information about the call, and then enter our main event loop. Note these two lines:

```
my $stay_connected = 1;
while ( $stay_connected ) {
```

The `$stay_connected` variable is simply a flag, and as long as it evaluates to true, the event loop keeps running. The script then polls the event socket for events:

```
my $e = $con->recvEventTimed(30);
```

The argument to `recvEventTimed` is the number of milliseconds to block while waiting for an event. The `$e` variable will evaluate to false if there are no events waiting:

```
if ( $e ) {
    ...
} else {
    # do other things while waiting for events...
}
```

The `else` block of this `if` statement can be used to let your code handle other operations while you are waiting for events to come. If an event does come in, we have this `if` block for checking the type of that event:

```
my $ev_name = $e->getHeader("Event-Name");
if ( $ev_name eq 'BACKGROUND_JOB' ) {
   my $call_result = $e->getBody();
   print "Result of call to $target was $call_result\n\n";
} elsif ( $ev_name eq 'DTMF' ) {
   my $digit = $e->getHeader("DTMF-Digit");
   print "Received DTMF digit: $digit\n";
   if ( $digit =~ m/\D/ ) {
      print "Exiting...\n";
      $stay_connected = 0;
   }
} else {
   # Some other event
}
```

We examine the event name for BACKGROUND_JOB or DTMF in the if and `elsif` checks (highlighted). We also have a bare `else` block, where we can handle events of other types, if we choose to do so. When we receive our BACKGROUND_JOB event, we display the result of the `originate` command. The rest of the script is spent in the event loop waiting for DTMF events. When a DTMF event comes in, we display the key that the caller pressed. If the key is not a digit (* or #), then the script will exit, otherwise the event loop will keep on processing. Note that we explicitly hang up the channel using the `uuid_kill` command.

There's more...

You can learn more about the mechanics of using `bgapi` by issuing some simple commands at `fs_cli`. Open an `fs_cli` session and try these commands:

```
/log 0
bgapi status
```

You will see a reply, something similar to the following:

```
+OK Job-UUID: f719939a-ffa1-49ca-a8b6-7f080febc2dc
```

You can manually watch for BACKGROUND_JOB events with this fs_cli command:

```
/event plain background_job
```

Now issue another bgapi status command. In addition to the reply, you will also see the actual BACKGROUND_JOB event. An abbreviated event looks like this:

```
Event-Name: [BACKGROUND_JOB]
. . .
Job-UUID: [f719939a-ffa1-49ca-a8b6-7f080febc2dc]
Job-Command: [status]
Content-Length: [177]
Content-Length: 177

UP 0 years, 0 days, 0 hours, 15 minutes, 2 seconds, 165 milliseconds,
501 microseconds
1 session(s) since startup
0 session(s) 0/90
1000 session(s) max
min idle cpu 0.00/100.00
```

The status command returns the BACKGROUND_JOB event immediately. However, the originate command will not return a BACKGROUND_JOB event until the originate API has succeeded (the call is answered) or failed (busy, no answer, and so on). Try it with your phone:

bgapi originate user/1000 9664

Replace 1000 with the extension number of your phone. You will get the +OK reply immediately, but you won't get the BACKGROUND_JOB event until the call is answered or goes to voicemail. One thing to keep in mind is that by default, if the far end sends back **early media**, then the originate is considered successful, even if that early media is a busy signal, special information tone (SIT), or a ring with no answer. To avoid considering early media as a success, use this:

```
bgapi originate {ignore_early_media=true}user/1000 9664
```

See also

▸ The *Setting up the event socket library*, *Establishing an inbound event socket connection*, and *Getting familiar with the fs_cli interface* recipes in this chapter

Using the ESL connection object for call control

Sometimes, it is convenient (or even necessary) to control a call from a script. In such cases, you can use the ESL connection object to control a call from an ESL script. This recipe will demonstrate a simple script that will answer a call, play a prompt, accept some caller input, and then route the call based on that input. With these basic concepts demonstrated, you will be able to write custom scripts that meet your specific needs.

Getting ready

This recipe is an example of an "outbound" connection from the FreeSWITCH dialplan to an ESL script. As such, you should have read the *Establishing an outbound event socket connection* recipe earlier in this chapter. This recipe will require at least two terminal windows: one for `fs_cli` and one for the script. Although the script presented here is written in Perl, the connection object applies to all ESL-enabled languages.

How to do it...

First, add a new extension to your dialplan by following these steps:

1. Edit or create a new file in `conf/dialplan/default/` named `02_event_socket.xml`.

2. Add this extension to the new file:

   ```xml
   <extension name="ESL Con Obj Example">
     <condition field="destination_number" expression="^(9960)$">
       <action application="log"
               data="INFO Starting ESL connection object example"/>
       <action application="socket"
               data="127.0.0.1:8040 sync full"/>
     </condition>
   </extension>
   ```

3. Save the file, exit, and then issue `reloadxml` at the `fs_cli` prompt.

   This extension will call your event socket script when the user dials 9960. Create the following script, or download it from the Packt Publishing website.

4. Create a new file called con_obj_example.pl in scripts/.

5. Add these lines to it:

```perl
#!/usr/bin/perl
use strict;
use warnings;
require ESL;
use IO::Socket::INET;
my $ip = "127.0.0.1";
my $sock = new IO::Socket::INET ( LocalHost => $ip,
                                  LocalPort => '8040',
                                  Proto => 'tcp',
                                  Listen => 1,
                                  Reuse => 1 );
die "Could not create socket: $!\n" unless $sock;
for(;;) {
  my $new_sock = $sock->accept();
  my $pid = fork();
  if ($pid > 0) {
    close($new_sock);
    next;
  } elsif ( $pid == 0 ) {
    my $host = $new_sock->sockhost();
    my $fd = fileno($new_sock);
    my $con = new ESL::ESLconnection($fd);
    my $info = $con->getInfo();
    my $uuid = $info->getHeader("unique-id");
    my $prompt = 'file_string://voicemail/vm-to_exit.wav';
    $prompt .= '!voicemail/vm-press.wav!digits/9.wav';
    $prompt .= ' ivr/ivr-that_was_an_invalid_entry.wav';
    $con->execute("answer");
    $con->execute("playback",
                  "ivr/ivr-welcome_to_freeswitch.wav");
    my $digits = "1";
    while($con->connected()) {
      while ( $digits != "9" && $con->connected() ) {
        $con->execute("play_and_get_digits",
                      "1 1 3 5000 # $prompt mydigits \\d+");
        my $e = $con->api("uuid_getvar","$uuid mydigits");
        $digits = $e->getBody();
```

```perl
        print "Received digit $digits\n";
        $con->execute("sleep","1000");
        $con->execute("playback","ivr/ivr-you_entered.wav");
        $con->execute("say","en number pronounced $digits");
        $con->execute("sleep","1000");
        if ( $digits == "9" ) {
          $con->execute("playback","voicemail/vm-goodbye.wav");
        }
      }
      $con->execute("hangup");
    }
    close($new_sock);
    exit(0);
  } else {
    die "Error forking new process: $!\n";
  }
}
```

6. Save the file and exit.

7. Make the script executable:

   chmod +x con_obj_example.pl

8. Run the script with this command:

   ./con_obj_example.pl

9. Once the script is running, dial 9960 and follow the voice prompts.

How it works...

This script runs constantly—a daemon in Unix parlance—and waits for socket connections from FreeSWITCH on port 8040. As soon as a socket connection is established, the script *forks a child process*. This child process then creates the $con ESL connection object. Once the $con object is created, we say a greeting to the caller and then enter the outer while loop. This loop causes the script to exit if the caller hangs up. The inner while loop uses the play_and_get_digits application to actually play the prompt and collect the digits from the caller. We then read back to the caller the digit they pressed using the say application. Finally, if the caller dialed the digit 9, then we say "Goodbye" and hang up. The child process then exits, but the parent (the daemon) is still running. You can have multiple simultaneous calls existing, and each one of them will get its own process.

You can use this script as a template to create your own interactive dialogs. All the caller interactions take place within the inner `while` loop, so focus your attention there. Also, if you plan to play various sound prompts to the caller, be sure to refer to the *Use phrase macros to build sound prompts* recipe in *Chapter 5, PBX Functionality*.

See also

▶   The *Setting up the event socket library, Establishing an outbound event socket connection*, and *Using fs_ivrd to manage outbound connections* recipes in this chapter

Using the built-in web interface

FreeSWITCH comes with a built-in web interface. It is made available by `mod_xml_rpc`, which is not loaded by default and, therefore, goes unnoticed sometimes.

Getting ready

You will need to make sure that `mod_xml_rpc` is built and loaded before trying to connect to the web interface. The `mod_xml_rpc` module is already compiled when using the Visual Studio 2008/2010 solution files with the FreeSWITCH source code. Linux and Mac OS X users will need to enable `mod_xml_rpc` in their FreeSWITCH installation. Follow these steps:

1.   Open `modules.conf` in the FreeSWITCH source directory, and remove the comment from the `#xml_int/mod_xml_rpc` line. Save the file and exit.

2.   Compile `mod_xml_rpc` with this command:

```
make mod_xml_rpc-install
```

3.   If you want to have `mod_xml_rpc` load automatically each time you start FreeSWITCH, then edit `conf/autoload_configs/modules.conf.xml` and uncomment this line:

```
<!-- <load module=" mod_xml_rpc "/> -->
```

Save the file and exit.

4.   If you do not want to load `mod_xml_rpc` automatically, then simply load it with this command from `fs_cli`:

```
load mod_xml_rpc
```

Once `mod_xml_rpc` is loaded, you are ready to start using the web interface.

How to do it...

Follow these steps:

1. Connect to the web interface with a browser by opening a URL such as
    `http://x.x.x.x:8080`, where `x.x.x.x` is the IP address of your
    FreeSWITCH server.

    By default, the interface uses port 8080. When the server asks for a username
    and password, enter "freeswitch" and "works" respectively. You will see a simple
    page displayed, like this:

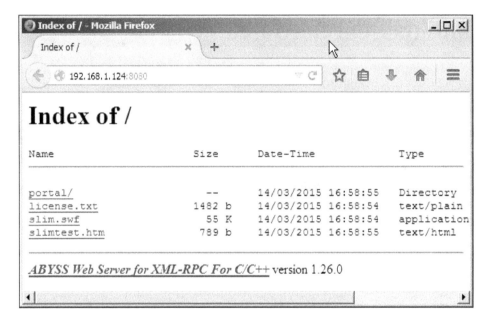

The files listed here are for the included Adobe Flash-based media player, which lets
you listen to audio sound files right from your browser, and are not of particular note.

2. Let's send a simple command to FreeSWITCH. The syntax for sending commands is
    `http://x.x.x.x:8080/webapi/cmd?args`, where `x.x.x.x` is the IP address of
    FreeSWITCH, `cmd` is the API command to send, and `args` represents any arguments
    to the command. Assuming that your IP address is 127.0.0.1, you can get the status
    of FreeSWITCH with the URL as `http://127.0.0.1:8080/webapi/status`.

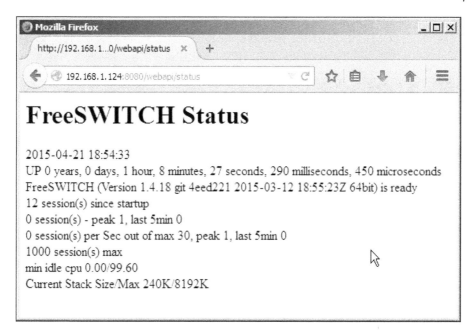

3. To view calls in progress, use `http://127.0.0.1:8080/webapi/show?channels`.

   Any API command that you can type at `fs_cli` can also be sent via the web interface.

How it works...

FreeSWITCH features a clever design that anticipates the possibility that commands
have been issued from the web-based interface instead of the console or `fs_cli` utility.
Commands that are "web aware" will respond with HTML-formatted data. For example, the
`help` command will respond with formatted output. Try sending this command from your
browser: `http://127.0.0.1:8080/webapi/help`. Notice the table and alternating
background colors. The `help` command is one of these web-aware commands. Note that
not all commands are like this, so if you issue a command and the response does not seem
formatted properly, then try the `api` or `txtapi` alternatives (the `api` method uses some
formatting for the output, whereas `txtapi` simply does a raw text dump for the output).

To get a better idea of the differences, issue each of these commands and see the response: `http://127.0.0.1:8080/api/help` and `http://127.0.0.1:8080/txtapi/help`. You have a number of options for sending and receiving data using the built-in web server.

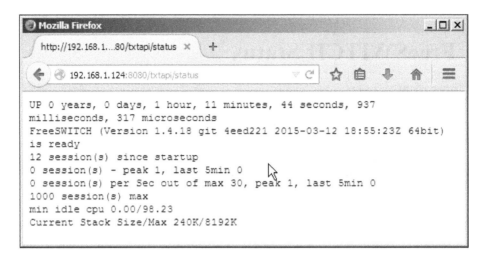

 Be sure to change the default username and password before putting this feature into production. Look for the `auth-user` and `auth-pass` parameters in `conf/autoload_configs/xml_rpc.conf.xml`.

There's more...

The built-in web server is used for several interesting features. The "XML RPC" in `mod_xml_rpc`.

This recipe focused entirely on using a web browser to communicate with FreeSWITCH. However, it is entirely possible to use traditional XML RPC clients in various programming languages. If you are familiar with XML RPC programming, then we recommend that you visit `http://wiki.freeswitch.org/wiki/Freeswitch_XML-RPC` to see some specific examples on using XML RPC. There is even an example for Drupal!

See also

 ▸   The *Accessing voicemail* recipe in *Chapter 5, PBX Functionality*

5
PBX Functionality

In this chapter, we will cover the following recipes:

- ▶ Creating users
- ▶ Accessing voicemail
- ▶ The company directory
- ▶ Using phrase macros to build sound prompts
- ▶ Creating XML IVR menus
- ▶ Music on hold
- ▶ Creating conferences
- ▶ Sending faxes
- ▶ Receiving faxes
- ▶ Basic text-to-speech with `mod_flite`
- ▶ Advanced text-to-speech with `mod_tts_commandline`
- ▶ Recording calls

Introduction

FreeSWITCH supports many features that are typically associated with a telephone system or **Private Branch Exchange** (**PBX**). The recipes in this chapter focus on a number of functions that are widely used in PBX systems, such as voicemail, faxing, call recording, IVR menus, and more.

Creating users

Each FreeSWITCH system has a directory of users. In most cases, a user is literally a person who has a telephone. When we say that we are "adding a user," we mean that we are creating a user account in the directory of users. Each "user" has the SIP credentials for making outbound calls, as well as a PIN number for accessing the voicemail. In fact, you cannot have a voicemail box without having a corresponding user account.

Getting ready

As a minimum, you will need a terminal window to issue commands to your system. To use the `add_user` script, your system will need to have Perl installed.

How to do it...

There are two basic steps for creating a user. The steps are as follows:

1. Add the user to the directory.
2. Add the corresponding extension number to the dialplan.

Let's assume we have a fresh installation of FreeSWITCH, which means that we have user ID's 1000 through 1019 (the `Local_Extension` in `conf/dialplan/default.xml` is set to route calls to these ID's).

Let's add a new user with these steps:

1. Open your terminal and perform `cd` to your FreeSWITCH source directory.
2. Linux users have to issue this command: `./scripts/perl/add_user 1020`.
3. Windows users have to use `perl scripts\perl\add_user 1020`.

```
root@vz124:~# cd /usr/src/freeswitch.git/
root@vz124:/usr/src/freeswitch.git# ./scripts/perl/add_user 1020

Added 1020 in file /usr/local/freeswitch/conf/directory/default/1020.xml

Operation complete. 1 user added.
Be sure to reloadxml.

If CPAN module Regexp::Assemble were installed this program would be able to sug
gest a regex for your new users.
root@vz124:/usr/src/freeswitch.git#
```

You should see some output confirming that the new user has been created. Next, we need to modify the `Local_Extension` in the `default` context. Perform these steps:

1. Open `conf/dialplan/default.xml` in a text editor.

2. Locate the dialplan extension named `Local_Extension`.

3. Change the expression from `^(10[01][0-9])$` to `^(10[012][0-9])$`.

4. Save the file and exit. Then issue a `reloadxml` command from `fs_cli`.

```
<extension name="Local_Extension">
    <condition field="destination_number" expression="^(10[012][0-9])$">
        <action application="export" data="dialed_extension=$1"/>
        <!-- bind_meta_app can have these args <key> [a|b|ab] [a|b|o|s] <app> --
>
        <action application="bind_meta_app" data="1 b s execute_extension::dx XM
L features"/>
        <action application="bind_meta_app" data="2 b s record_session::$${recor
dings_dir}/${caller_id_number}.${strftime(%Y-%m-%d-%H-%M-%S)}.wav"/>
                                                           272,64          33%
```

User 1020 is now ready for use. To test, have a SIP phone register as user "1020" and then call it from another phone.

How it works...

The `add_user` script simply adds a new file to the directory. In the case of user ID 1020, it created the `conf/directory/default/1020.xml` file. Once that file is created (and you have issued a `reloadxml` command from `fs_cli`), a SIP phone can register as that user and make calls. However, the dialplan isn't set up to handle someone dialing 1020 by default, which is why we had to update the `Local_Extension` in `default.xml`. The default `<condition>` for `Local_Extension` is as follows:

```
<condition field="destination_number"
        expression="^(10[01][0-9])$">
```

This pattern matches 1000, 1001, and so on up to 1019. We changed the `<condition>` line to read as follows:

```
<condition field="destination_number"
        expression="^(10[012][0-9])$">
```

Our new pattern adds 1020, 1021, and so on up to 1029 to `Local_Extension`. But why the entire range instead of just "1020"? There is nothing preventing you from doing this, but it is quite common to add users in blocks and not one at a time. If you prefer, you can use the following pattern:

```
<condition field="destination_number"
        expression="^(10[01][0-9]|1020)$">
```

However, as mentioned, this means that if you want to add user 1021, then you will need to come back and change this regular expression pattern again.

There's more...

The `add_user` script has many useful features (run `add_user --help` to see the full set of options). One such feature is adding a block of users. For example, if we want to complete the block of 1020, 1021, and so on up to 1029, we need not run the script for each user to add. Instead, we specify a range with the `--users` argument:

```
./scripts/perl/add_user --users=1020-1029
```

Note that the `add_user` script will not overwrite existing users.

See also

 ▸ Refer to the *Configuring an SIP phone to register with FreeSWITCH* recipe in *Chapter 2, Connecting Telephones and Service Providers*

Accessing voicemail

Voicemail is a very common feature of PBX systems. This recipe shows how to access the voicemail for a user.

Getting ready

You will need at least one telephone registered on your system, though it is easier to test with two or more phones. Have another user call your registered extension. The destination extension should let the call go to the voicemail. Also, the caller should leave a message and hang up. Once a message is left, the target phone can access the voicemail.

How to do it...

The simplest way to access the voicemail is to simply dial *98 from the phone. The system will ask for the user ID and then the password (by default, the password is the same as the user ID). Let's assume that user 1001 is checking their voicemail messages. Then they would follow these steps:

1. Dial *98 and wait for the system to answer.
2. Enter the ID and press # (`1001#` in our example).
3. Enter the password and press # (`1001#` in our example).

4. New messages are automatically played.

Simply hang up the phone to exit from voicemail.

How it works...

The voicemail system is really nothing more than a specific type of IVR system. In this case, the user can log in and has several choices. The main menu options are as follows:

Key	Action
1	Listen to new messages
2	Listen to saved messages
5	Advanced options
#	Exit the voicemail

While listening to new or saved messages, the user has these options:

Key	Action
1	Listen to the message from the beginning
2	Save the message
4	Rewind
6	Fast forward
7	Delete
0	Reread from the beginning
*	Skip the envelope information (date/time and sender)

After the message has been played, the options are as follows:

Key	Action
1	Listen to the message from the beginning
2	Save the message
7	Delete

Here are the advanced menu options:

Key	Action
1	Record a greeting
2	Choose a greeting
3	Record the name

Key	Action
6	Change the password
0	Main menu

Most users will find the FreeSWITCH voicemail system very familiar, as it is modeled on the voicemail systems used by most of the major mobile phone carriers.

See also

► Refer to the *Configuring a SIP phone to register with FreeSWITCH* recipe in *Chapter 2, Connecting Telephones and Service Providers*

The company directory

Most companies have some form of dial-by-name directory. This recipe will show you how to add a company directory to your FreeSWITCH installation using mod_directory.

How to do it...

Enable and build mod_directory by following these steps:

1. Open modules.conf in your FreeSWITCH source directory.

2. Uncomment this line:

   ```
   #applications/mod_directory
   ```

3. Save the file and exit.

4. Linux/Unix users have to issue the proper make command:

   make mod_directory-install

Allow mod_directory to be loaded when FreeSWITCH starts:

1. Open conf/autoload_configs/modules.conf.xml in a text editor.

2. Uncomment this line:
   ```
   <!--<load module="mod_directory"/>-->
   ```

3. Save the file and exit.

4. Restart FreeSWITCH.

5. Start fs_cli and issue the command show application.

You should see an application named `directory` in the list. Next, we need to add a simple dialplan extension that will let us test:

1. Open `conf/dialplan/default/01_Custom.xml` in a text editor.

2. Add these lines:

    ```
    <include>
      <extension name="dial by name">
        <condition field="destination_number"
              expression="^(1411)$">
          <action application="directory"
              data="default ${domain}"/>
        </condition>
      </extension>
    </include>
    ```

3. Save the file and exit.

```
<include>
  <extension name="dial by name">
    <condition field="destination_number" expression="^(1411)$">
      <action application="directory" data="default ${domain}"/>
    </condition>
  </extension>
</include>
~
~
~
<switch/conf/dialplan/default/01_Custom.xml" 7L, 217C written 7,1          All
```

The last thing to do is to make sure that at least one user in the directory has the `directory_full_name` or `effective_caller_id_name` variable set in the directory entry. For now, we will set `directory_full_name` on user 1000:

1. Open `conf/directory/default/1000.xml` in a text editor.

2. Add this line to the `<variables>` section:

    ```
    <variable name="directory_full_name" value="Ada Lovelace"/>
    ```

3. Save the file and exit. Issue the `reloadxml` command from `fs_cli`.

```
    <variable name="effective_caller_id_number" value="1000"/>
    <variable name="outbound_caller_id_name" value="$${outbound_caller_name}"/
>
    <variable name="outbound_caller_id_number" value="$${outbound_caller_id}"/
>
    <variable name="directory_full_name" value="Linda Lovelace"/>
    <variable name="callgroup" value="techsupport"/>
  </variables>
</user>
<reeswitch/conf/directory/default/1000.xml" 19L, 819C written 15,55        91%
```

At this point, you are ready to test. Dial *1411* from your phone and listen to the options. For this test, dial the numbers corresponding to the first three letters of the last name (*568* for "L-O-V"), and listen to the results.

How it works...

The `directory` application gets its information from the user directory. Using the `directory_full_name` variable, we specify the first and last names for the purpose of searching the user directory. You can also use the `effective_caller_id_name` variable if you wish. This variable controls the caller ID name displayed when the user makes outbound calls. If, for any reason, this is not the name you want searched, then use `directory_full_name`, which will always supersede `effective_caller_id_name` for dial-by-name searches.

Most likely, in your initial test, you did not hear someone's voice saying, "Ada Lovelace." Instead, you heard the system spelling out the name. This is how `mod_directory` handles the case where the user has not recorded their name. If you log in to the voicemail system and record a name prompt (option 5 from the VM main menu and then option 3), then the system will use that recording instead of spelling out the user's name.

There's more...

You have two parameters that you can set for each user to customize the behavior of the directory application:

- ▶ `directory-visible`: Set this parameter to `false` to prevent the user from being included in directory searches. This is useful for keeping the directory from being cluttered with entries such as "hallway phone", and "guest phone." It is also handy for keeping VIP extensions from being included.

- ▶ `directory-exten-visible`: Set this parameter to `false` to prevent the `directory` application from voicing the target user's extension number (some operations prefer to keep extension numbers from being public).

Both of the preceding parameters default to `true`, so keep that in mind when you are building your user database.

See also

- ▶ Refer to the *Accessing voicemail* and *Creating users* recipes in this chapter

Using phrase macros to build sound prompts

It is frequently necessary to piece together smaller sound recordings to create longer ones. The FreeSWITCH phrase macro system is a very powerful tool for not only piecing together individual sound files, but also for adding a bit of logic so that your phrases are more than mere amalgamations of individual sound prompts.

In this recipe we will create a simple dialplan extension that will read back to the caller their extension number. We will use a phrase macro to handle the work of stitching together sound prompts and utilizing the `say` application to read back the caller's extension number.

Getting ready

You will need a text editor and at least one phone for this recipe. It is also recommended that you review the phrase file for your language. For English, this can be found in the FreeSWITCH source directory in `docs/phrase/phrase_en.xml`. The `phrase_en.xml` file contains both the filename of each prerecorded prompt as well as the actual spoken text. Prompts are divided into sections such as `voicemail`, `ivr`, `currency`, `digits`, and `time`. By far, the largest collection of sound prompts is in the `ivr` section.

How to do it...

Start by adding the extension to the dialplan:

1. Create or edit the `conf/dialplan/default/02_Custom.xml` file.
2. Add these lines:

```xml
<include>
<extension name="who's calling">
  <condition field="destination_number"
    expression="^(1500)$">
    <action application="answer"/>
    <action application="playback"
            data="phrase:whoami:${username}"/>
    <action application="hangup"/>
  </condition>
</extension>
</include>
```

3. Save the file and exit.

```
<include>
<extension name="who's calling">
  <condition field="destination_number" expression="^(1500)$">
    <action application="answer"/>
    <action application="playback"
          data="phrase:whoami:${username}"/>
    <action application="hangup"/>
  </condition>
</extension>
</include>
~
<conf/dialplan/default/02_Custom.xml" [New] 10L, 297C written 1,1        All
```

Next, create the phrase macro:

1. Create or edit the `conf/lang/en/ivr/custom.xml` file.

2. Add the following lines:

```
<macro name="whoami">
  <input pattern="^(\d+)$">
    <match>
      <action function="play-file"
              data="ivr/ivr-extension_number.wav"/>
      <action function="sleep"
              data="100"/>
      <action function="say"
              data="$1"
              method="pronounced"
              type="number"/>
    </match>
    <nomatch>
      <action function="play-file"
          data="ivr/ivr-that_was_an_invalid_entry.wav"/>
    </nomatch>
  </input>
</macro>
```

3. Save and exit.

4. Issue the `reloadxml` command at `fs_cli`.

Test the new extension by dialing *1500*.

```
<macro name="whoami">
  <input pattern="^(\d+)$">
    <match>
      <action function="play-file"
              data="ivr/ivr-extension_number.wav"/>
      <action function="sleep"
              data="100"/>
      <action function="say"
              data="$1"
<eswitch/conf/lang/en/ivr/custom.xml" [New] 19L, 496C written 5,15        Top
```

How it works...

The key to this operation is this line in the dialplan extension we created:

```
<action application="playback"
    data="phrase:whoami:${username}"/>
```

The `playback` application normally takes a filename as an argument. However, if the argument begins with `phrase:`, then `playback` will look for a phrase macro instead of an audio file. In this case, we call a phrase macro named `whoami` and give it the argument of `${username}`, which contains the ID of the calling user. At this point, the phrase macro takes control.

The argument passed to the macro gets handled with this line:

```
<input pattern="^(\d+)$">
```

The input value is matched against the regular expression in the `pattern` option. If `${username}` contains only digits, our pattern will capture those into the `$1` special variable. After the regular expression we have a bit of logic to help us decide what to do. If the input matches the pattern, then the actions inside the `<match>` node will be executed. If there is not a match because `${username}` contains something else (spaces, letters, and so on) besides digits, then the actions inside the `<nomatch>` node will be executed (we simply play a message that says "That was an invalid entry").

You have probably figured out by now that the actions contained inside `match` (or `nomatch`) are executed sequentially. You can also see that phrase macros are not limited to playing individual sound files. You can call functions such as `sleep` and `say` to customize the way the prompt is played to the user. You can even call a text-to-speech application, if you have one installed.

There's more...

It is possible to execute many different operations using phrase macros. In fact, the FreeSWITCH voicemail system uses phrase macros extensively. Look through `conf/lang/en/vm/sounds.xml` to see all the different phrase macros that `mod_voicemail` uses. Keep in mind that you can use any of the phrase macros in `sounds.xml` as long as you call them with the correct arguments.

One particularly useful phrase macro is called `voicemail_record_file_check`. Consider the case where you have a custom application where you are asking the caller to record a prompt. This macro allows you to have a custom phrase that says something like "press 1 to listen, press 2 to save, press 3 to rerecord." As an example, you can use `play_and_get_digits` to tell the caller what to do:

```
<action application="play_and_get_digits" data="1 1 3 4000 #
phrase:voicemail_record_file_check:1:2:3 ivr/ivr-invalid_entry.wav
selection \d"/>
```

The preceding action will tell the caller: "Press 1 to listen to the recording; press 2 to save the recording; press 3 to rerecord." It will then capture the input into the `${selection}` channel variable. Note that the options voiced to the caller are customizable with this macro. Calling the macro with `phrase:voicemail_record_file_check:4:5:6` will tell the caller: "Press 4 to listen to the recording; press 5 to save the recording; press 6 to rerecord".

A good way to learn more about phrase macros is to enter at the FreeSWITCH console `fsctl loglevel 7`, then enter `console loglevel 7`, and watch the console while calling to the voicemail. You will be able to see in real time how FreeSWITCH parses the phrase macro and performs the actions therein.

```
freeswitch@internal> fsctl loglevel 7
+OK log level: DEBUG [7]

freeswitch@internal> console loglevel 7
+OK log level 7 [7]
+OK console log level set to DEBUG

freeswitch@internal>
```

See also

- Refer to the *Basic text-to-speech with mod_flite* and *Advanced text-to-speech with mod_tts_commandline* recipes later in this chapter

Creating XML IVR menus

FreeSWITCH has a simple but flexible system for building IVR-style menus for caller interaction. In this recipe, we will create a custom menu that is very similar to the demo IVR that is part of the default FreeSWITCH configuration.

Getting ready

You will need a text editor and a telephone for testing. We will create a custom menu for extension number 5002, and use a generic greeting that comes with the FreeSWITCH sound files. To use the `dial-by-name` directory, be sure to complete the *The company directory* recipe covered earlier in this chapter.

How to do it...

Create the menu definition by following these steps:

1. Open a text editor and create a new file called `conf/ivr_menus/custom_ivr.xml`.
2. Add these lines:

```xml
<menu name="simple_greeting"
      greet-long="ivr/ivr-generic_greeting.wav"
      greet-short="ivr/ivr-generic_greeting.wav"
      invalid-sound="ivr/ivr-that_was_an_invalid_entry.wav"
      exit-sound="voicemail/vm-goodbye.wav"
      confirm-attempts="3"
      timeout="10000"
      inter-digit-timeout="2000"
      max-failures="3"
      max-timeouts="3"
      digit-len="4">
  <entry action="menu-exec-app" digits="9"
        param="directory default ${domain}"/>
  <entry action="menu-exec-app"
        digits="/^(10[01][0-9])$/"
        param="transfer $1 XML features"/>
  <entry action="menu-top" digits="*"/>
</menu>
```

3. Save the file.

Next, we create a simple extension that lets us test our menu:

1. Open `conf/dialplan/default/03_Custom.xml` in a text editor.
2. Add this extension:

```xml
<include>
<extension name="sample greeting">
  <!-- Good morning 12am to 11:59 -->
  <condition hour="0-11" break="never">
    <action application="set" data="tod=morning"
        inline="true"/>
  </condition>
  <!-- Good afternoon 12pm to 17:59 -->
  <condition hour="12-17" break="never">
    <action application="set" data="tod=afternoon"
        inline="true"/>
  </condition>
  <!-- Good morning 18:00 to 23:59 -->
  <condition hour="18-23" break="never">
    <action application="set" data="tod=evening"
        inline="true"/>
  </condition>
  <condition field="destination_number"
        expression="^5002$">
    <action application="answer"/>
    <action application="sleep" data="1000"/>
    <action application="playback"
            data="ivr/ivr-good_${tod}.wav"/>
    <action application="sleep" data="500"/>
    <action application="ivr" data="simple_greeting"/>
  </condition>
</extension>

</include>
```

3. Save and exit.
4. Issue the `reloadxml` command at `fs_cli`.

Test your new extension by dialing *5002*.

There's more...

Often, it is beneficial to use a phrase macro with an IVR menu. For example, in our dialplan, we manually compute the time of day and voice it to the caller. We then launch the `ivr` application with our generic greeting. This is not optimal for a few reasons. First off, having a simple `.wav` file for our greeting means that we are stuck with whatever is recorded. Secondly, using a phrase macro gives us a bit more flexibility in how we use our macros. Let's improve our menu by using a phrase macro. Our goals will be as follows:

- Add "To repeat these options, press *" to our greeting
- Skip "Good morning/afternoon/evening" when repeating our options
- Clean up the readability of our dialplan

As you will see, using a phrase macro accomplishes all of this, and more. First, let's clean up the dialplan. We open `conf/dialplan/default/03_Custom.xml` and edit our extension so that it has only these lines:

```
<extension name="sample greeting">
  <condition field="destination_number" expression="^5002$">
    <action application="answer"/>
    <action application="ivr" data="simple_greeting"/>
  </condition>
</extension>
```

Now let's create a separate extension that always gets executed at the beginning of the dialplan. Normally you do this at the beginning of the `default` context. Open `conf/dialplan/default.xml` and add this as the first extension in the default context:

```
<extension name="set_tod" continue="true">
  <!-- Good morning 12am to 11:59 -->
  <condition hour="0-11" break="never">
    <action application="set"
            data="tod=morning"
            inline="true"/>
  </condition>
  <!-- Good afternoon 12pm to 17:59 -->
  <condition hour="12-17" break="never">
    <action application="set"
            data="tod=afternoon"
            inline="true"/>
  </condition>
  <!-- Good morning 18:00 to 23:59 -->
  <condition hour="18-23" break="never">
    <action application="set"
            data="tod=evening"
```

```
                        inline="true"/>
        </condition>
    </extension>
```

Adding this extension to the dialplan allows all calls in the `default` context to have the `tod` channel variable set. This in turn lets any extension (or script, or phrase macro) get access to `tod`, not just our custom extension.

Next, we open `conf/ivr_menus/custom_ivr.xml` and change these two lines to use our macro:

```
        greet-long="phrase:simple_greeting:long"
        greet-short="phrase:simple_greeting:short"
```

Finally, add the new macro. It's a bit long; however, it accomplishes a lot for us. Open `conf/lang/en/ivr/custom.xml` and add a new macro:

```
    <macro name="simple_greeting">
      <input pattern="^(long)$" break-on-match="true">
        <match>
          <action function="sleep"
                  data="1000"/>
          <action function="play-file"
                  data="ivr/ivr-good_${tod}.wav"/>
          <action function="sleep"
                  data="500"/>
          <action function="play-file"
                  data="ivr/ivr-generic_greeting.wav"/>
          <action function="sleep"
                  data="500"/>
          <action function="play-file"
                  data="ivr/ivr-to_repeat_these_options.wav"/>
          <action function="sleep"
                  data="250"/>
          <action function="play-file"
                  data="voicemail/vm-press.wav"/>
          <action function="sleep"
                  data="100"/>
          <action function="play-file"
                  data="ascii/42.wav"/>
        </match>
      </input>
      <input pattern="^(short)$">
        <match>
          <action function="play-file"
                  data="ivr/ivr-generic_greeting.wav"/>
```

```
    <action function="sleep"
            data="500"/>
    <action function="play-file"
            data="ivr/ivr-to_repeat_these_options.wav"/>
    <action function="sleep"
            data="250"/>
    <action function="play-file"
            data="voicemail/vm-press.wav"/>
    <action function="sleep"
            data="100"/>
    <action function="play-file"
            data="ascii/42.wav"/>
  </match>
 </input>
</macro>
```

After saving all the files, issue the `reloadxml` command from `fs_cli`. Try calling 5002, and this time, press * to repeat the options. Upon repetition, the system will not say "Good morning", and so on. In addition to being more functional, the phrase macro method also makes it easier for you to make changes to the greeting that you play to your callers.

See also

▶ Refer to the *The company directory* recipe earlier in this chapter

Music on hold

Music on hold (**MOH**) is a common feature of modern phone systems. FreeSWITCH allows you to have many different MOH selections.

Getting ready

You will need some music files if you wish to customize the MOH. Also, if you have MP3 files that you would like to use for MOH, then you will need a utility that can convert them into standard WAV files. A freely available tool can be found at `http://www.mpg123.de`. You will also need a text editor and a telephone connected to your FreeSWITCH server.

How to do it...

The first thing to do is to install the default MOH files from the FreeSWITCH download site. Linux/Unix users can issue the following command from the FreeSWITCH source directory:

```
make cd-moh-install
```

On Windows, the sound files are installed automatically as part of the MSVC solution file.

Once the sounds are installed, you can confirm that they work by dialing *9664* (no `reloadxml` or system restart is necessary).

How it works...

The `make` command you just issued installs the MOH files in 8 kHz, 16 kHz, 32 kHz, and 48 kHz sampling rates (the Windows build automatically installs these as well). The default dialplan extension number 9664 (9MOH) will play the default music on hold files to the caller. The music is supplied by the `mod_local_stream` module. It is possible to customize the MOH on your system by adding other streams.

There's more...

Let's create an alternative MOH source and test it out. If you have a few MP3 or WAV files that you would like to use, then be ready to copy them to a new subdirectory on the FreeSWITCH server. In this example, we will download a few pieces of royalty-free music, along with an attribution sound clip, and then we will convert them into WAV files using the `mpg123` tool.

The `mpg123` tool can also be built automatically as part of `mod_shout`:

cd /usr/src/freeswitch

make mod_shout-install

/usr/src/freeswitch/libs/mpg123-1.13.2/src/mpg123

We start by creating a directory for our new sounds. In Linux/Unix, do this:

mkdir /usr/local/freeswitch/sounds/music/custom1

cd /usr/local/freeswitch/sounds/music/custom1

Copy your MP3 files to this directory. Alternatively, you can download some royalty-free music such as the following:

```
wget http://incompetech.com/music/royalty-free/mp3-royaltyfree/Skye%20
Cuillin.mp3
wget http://incompetech.com/music/royalty-free/mp3-royaltyfree/
Parisian.mp3
wget http://incompetech.com/music/royalty-free/mp3-royaltyfree/
credits%20sounder.mp3
```

```
root@vz124:~# wget http://incompetech.com/music/royalty-free/mp3-royaltyfree/Par
isian.mp3
--2015-05-29 00:52:40--  http://incompetech.com/music/royalty-free/mp3-royaltyfr
ee/Parisian.mp3
Resolving incompetech.com (incompetech.com)... 76.72.166.146
Connecting to incompetech.com (incompetech.com)|76.72.166.146|:80... connected.
HTTP request sent, awaiting response... 200 OK
Length: 1753171 (1.7M) [application/octet-stream]
Saving to: `Parisian.mp3'

100%[===================================>] 1,753,171   1.55M/s   in 1.1s

2015-05-29 00:52:41 (1.55 MB/s) - `Parisian.mp3' saved [1753171/1753171]

root@vz124:~# ....
```

Now convert your MP3 files into WAV files and remove the MP3 files:

```
for i in *.mp3; do mpg123 -m -r 8000 -w "`basename "$i" .mp3`".wav "$i";
done

rm *.mp3
```

You now have a set of 8 kHz WAV files that can be used as a music source. The next step is to create the actual file stream.

Open `conf/autoload_configs/local_stream.conf.xml` and add this new stream definition:

```xml
<directory name="custom1" path="$${sounds_dir}/music/custom1">
  <param name="rate" value="8000"/>
  <param name="shuffle" value="true"/>
  <param name="channels" value="1"/>
  <param name="interval" value="20"/>
  <param name="timer-name" value="soft"/>
</directory>
```

Save the file and close. Open `conf/dialplan/default/04_Custom.xml` and add this extension:

```xml
<include>
<extension name="hold_music">
  <condition field="destination_number" expression="^96642$">
    <action application="playback" data="${custom1}"/>
  </condition>
</extension>
</include>
```

Save the file and close. Finally, we need to create the ${custom1} global variable that can be used wherever we want to play our custom MOH. Open conf/vars.xml in a text editor and add this line:

```
<X-PRE-PROCESS cmd="set"
data="custom1=local_stream://custom1"/>
```

Save the file and exit.

Because we changed an X-PRE_PROCESS directive, 'which is read/executed once at FreeSWITCH startup, we need to restart FreeSWITCH.

When the module is reloaded, issue this command:

show_local_stream

Among the local streams listed, there should be your new custom1 stream:

```
custom1,/usr/local/freeswitch/sounds/music/custom1
```

Now you can dial 96642, and you should hear your new music source.

Now you can also use ${custom1} as the source of MOH and as a sound for ringback and transfer ringback operations.

Creating conferences

FreeSWITCH excels at letting multiple parties connect to a single conference "room" where they can all hear and speak to one another. The default configuration has some examples of conferences that we can use as a starting point. Keep in mind that in FreeSWITCH, there is no need explicitly to "create" a conference room—the conference dialplan application does all the work for us.

Getting ready

In addition to a text editor, you will need at least two phones for testing, and preferably another person or two so that you can verify that your conference rooms are working. Also, make sure that you have the default FreeSWITCH configuration installed and the sound and music files added.

How to do it...

Follow these steps:

1. Dial 3000 and listen. You will be put into a standard conference room, and if you are the only person there, then after the announcement, you will hear hold music.

2. Dial *3000* from another phone, and both persons are in the same conference.

3. Add more parties by dialing *3000* from other phones.

How it works...

The default FreeSWITCH dialplan has conferences predefined and ready for use (note that the conferences are not actually "active" until at least one person calls). The default dialplan has these conference extensions:

Extension range	Conference audio sampling rate
3000-3099	8 kHz
3100-3199	16 kHz
3200-3299	32 kHz
3300-3399	48 kHz

The sampling rate is the maximum sampling rate for all members. As an example, if you have a phone that uses G.722 at 16 kHz and you call 3000, then your audio will be resampled to 8 kHz before being sent out to the other participants. If you have multiple parties whose phones support wide-band audio, then be sure to use a conference room with a higher sampling rate to take advantage of the higher quality audio.

If you simply need to have several people, each hearing all others, in a conference room, then use the conference extensions in the default dialplan, and modify the extension numbers as needed.

There's more...

Conferences support many features, such as caller controls and moderators. Read on for information about using these other features.

Caller controls

There are many controls that you can give to callers in a conference. The most common ones are as follows:

▸ **Talk volume**: This is the volume of the audio that the caller sends (that is, gain control).

▸ **Listen volume**: This is the volume of the audio that the caller hears.

▸ **Energy threshold**: This is the minimum energy level of the audio from the caller required in order to be considered talking. Raising the energy level will cut down on background noise when a participant is in a noisy environment. For example, when FreeSWITCH "thinks" a person is not talking (sound from their microphone is below the threshold) they are muted, and are automatically unmuted when their sound goes above the threshold.

To see the default controls, open `conf/autoload_configs/conference.conf.xml` and locate the following section:

```xml
<caller-controls>
  <group name="default">
    <control action="mute" digits="0"/>
    <control action="deaf mute" digits="*"/>
    <control action="energy up" digits="9"/>
    <control action="energy equ" digits="8"/>
    <control action="energy dn" digits="7"/>
    <control action="vol talk up" digits="3"/>
    <control action="vol talk zero" digits="2"/>
    <control action="vol talk dn" digits="1"/>
    <control action="vol listen up" digits="6"/>
    <control action="vol listen zero" digits="5"/>
    <control action="vol listen dn" digits="4"/>
    <control action="hangup" digits="#"/>
  </group>
</caller-controls>
```

The name of this call control group is "default" and it cannot be modified (a "default" group is always needed). However, you can define your own custom caller controls groups and then add them to your conference definitions. Each conference is defined by a "profile" in the `<profiles>` section of `conference.conf.xml`. Let's say you created a caller control group named "custom." To set the conference profile to use those controls, just add this parameter to the profile:

```xml
<param name="caller-controls" value="custom"/>
```

Now, all callers who join this conference will have your custom caller controls.

Conference moderator and PIN

Some conferences have the concept of a "moderator" who has some level of control over the conference. In FreeSWITCH, the conference moderator is simply a conference member whose absence or presence can optionally affect the conference. There are primarily two ways by which the moderator affects the conference:

- All members wait until the moderator arrives
- The conference ends (all members are disconnected) when the moderator leaves

A moderator is created by modifying the conference application's argument in the dialplan. Compare these two lines:

```
<action application="conference" data="$1@default"/>
<action application="conference"
    data="$1@default+flags{moderator}"/>
```

Notice that we add `+flags{moderator}` to set the caller that comes from the extension that contains this `action` application as the moderator. You can have multiple flags separated by commas, for example, `+flags{moderator,mute}`.

Adding a PIN to the conference is simple as well. The same two conferences in the preceding code can have a PIN added, such as follows:

```
<action application="conference" data="$1@default+1234"/>
<action application="conference"
        data="$1@default+1234+flags{moderator}"/>
```

In both cases, the conference PIN is "1234," and the caller will not be allowed into the conference until they enter the correct PIN number.

Sending faxes

FreeSWITCH can transmit electronic documents to a destination fax machine. Only TIFF documents can be transmitted. However, it is possible to convert a number of formats (for example, PDF) to TIFF. This recipe will discuss some common and freely available tools.

Getting ready

In simple terms, sending a fax requires only a few things such as a TIFF file, gateway, and destination fax machine (for testing purposes, you can download a sample TIFF file from `http://files.freeswitch.org/txfax-sample.tiff`). Put your TIFF file into a known location. For our example, we will use `/tmp/txfax-sample.tiff`. The gateway is your connection to the PSTN, and the fax machine will simply be the device that answers your outbound phone call. Even, if you do not have a gateway or a fax machine handy, you can still try out this recipe by having FreeSWITCH send the fax to itself using the `fax_receive` extension in the default dialplan.

How to do it...

In most cases involving fax transmissions, you will be making an outbound call to a fax machine (the A leg) and then execute the `txfax` dialplan application. Execute these steps to send a simple fax transmission to FreeSWITCH itself:

1. Launch `fs_cli`.
2. Execute this command:

```
originate loopback/9178 &txfax(/tmp/txfax-sample.tiff)
```

Watch the console. Eventually, the fax transmission should be successfully completed.

How it works...

The `originate` command creates the outbound leg of the fax call. In this example, we are literally making a call within our own FreeSWITCH server using the loopback channel. The target extension is "9178." In a real example, we would, of course, be dialing an external number. For example, we can do this:

```
originate sofia/gateway/my_gw/18005551212 &txfax(/tmp/txfax-sample.tiff)
```

In any case, once the A leg is answered, the `txfax` application is called. If all goes well, the fax transmission should go through (in the case of transmission to 9178 to ourselves, a received file will be found in `/tmp/rxfax.tiff`).

There's more...

Faxing can be tricky. The following sections offer some helpful suggestions.

Diagnosing fax issues

Fax problems are quite common, especially in a VoIP environment. When a fax transmission fails for some reason, it helps to know what happened. If you are using XML CDRs, you will automatically have a number of channel variables populated on every fax call, whether successful or not. Here is a sample:

```
<fax_v17_disabled>0</fax_v17_disabled>
<fax_ecm_requested>1</fax_ecm_requested>
<fax_filename>/tmp/txfax.tif</fax_filename>
<fax_success>1</fax_success>
<fax_result_code>0</fax_result_code>
<fax_result_text>OK</fax_result_text>
<fax_ecm_used>on</fax_ecm_used>
```

```
<fax_local_station_id>SpanDSP%20Fax%20Ident</fax_local_station_id>
<fax_remote_station_id>SpanDSP%20Fax%20Ident</fax_remote_station_id>
<fax_document_transferred_pages>1</fax_document_transferred_pages>
<fax_document_total_pages>1</fax_document_total_pages>
<fax_image_resolution>8031x3850</fax_image_resolution>
<fax_image_size>24111</fax_image_size>
<fax_bad_rows>0</fax_bad_rows>
<fax_transfer_rate>14400</fax_transfer_rate>
```

Use this information to diagnose your fax issues. The `<fax_result_text>` is probably the most useful. It will report a description of error, if any.

Helpful software

There are numerous **Free and Open Source Software** (**FOSS**) packages that are available for help with handling PDF and TIFF files. Members of the FreeSWITCH community have had particular success with Ghostscript, which lets you convert to and from PDF and PostScript files.

A common operation is to convert a PDF file to TIFF before transmitting via fax. The following command will make a standard-resolution TIFF file from the source PDF:

```
gs -q -sDEVICE=tiffg3 -r204x98 -dBATCH -dPDFFitPage -dNOPAUSE
-sOutputFile=out.tif in.pdf
```

For a higher resolution file, use this command:

```
gs -q -sDEVICE=tiffg3 -r204x196 -dBATCH -dPDFFitPage -dNOPAUSE
-sOutputFile=out.tif in.pdf
```

All commands have been taken from `http://www.soft-switch.org/spandsp_faq/ar01s14.html`, a guide to converting `.pdf` to `.tiff` to be faxed. This page is written by Steve Underwood, the godfather of fax and DSP.

The Ghost Script executable (`gs`) is suited quite well for shell scripting.

See also

- Refer to the *Receiving faxes* recipe in this chapter
- Refer to the *Using XML CDRs* recipe in *Chapter 3, Processing Call Detail Records*

Receiving faxes

The preceding recipe described the process of sending a fax. This recipe will describe the process of receiving a fax.

Getting ready

In its simplest format, receiving a fax only requires that you route an incoming call to an extension, which then executes the `rxfax` dialplan application. As with the previous recipe, we can use our FreeSWITCH server to be both the sender and the receiver of the fax. For our test, we will use the same file we used in the *Sending faxes* recipe—/tmp/txfax-sample.tiff.

How to do it...

Execute these steps to carry out a simple fax transmission and reception:

1. Launch fs_cli.

2. Execute this command:

```
originate loopback/9178 &txfax(/tmp/txfax-sample.tiff)
```

Watch the console. Eventually, the fax transmission should be successfully completed.

How it works...

We use the `fax_receive` extension in the default dialplan to receive the fax transmission. This extension is quite simple:

```
<extension name="fax_receive">
  <condition field="destination_number" expression="^9178$">
    <action application="answer" />
    <action application="playback" data="silence_stream://2000"/>
    <action application="rxfax"
data="$${temp_dir}/rxfax.tif"/>
    <action application="hangup"/>
  </condition>
</extension>
```

The received fax is stored in `/tmp/rxfax.tif`. Feel free to modify the filename. For example, if you have a `faxes/` subdirectory off the main `freeswitch` install directory you can do this:

```
<action application="rxfax" data="${base_dir}/faxes/${uuid}.tif"/>
```

Each incoming fax will have a unique filename and be stored in the `faxes/` subdirectory.

There's more...

Receiving faxes is usually a part of a larger process or system. The following sections have some useful information about handling inbound fax transmissions.

Detecting inbound faxes

Let's say you have an automated attendant that answers all incoming calls and lets callers choose their destinations. Occasionally, a fax call may come in. Instead of disconnecting, you can detect the fax and send the call to a fax handler extension for processing.

This can be accomplished with the `spandsp_start_fax_detect` application. Consider this dialplan xml additional file:

```
<include>
  <extension name="fax detect test">
    <condition field="destination_number" expression="123456">
      <action application="answer"/>
      <action application="set"
          data="transfer_ringback=${us-ring}"/>}"/>}"/>}"/>
      <action application="spandsp_start_fax_detect"
          data="transfer '9178 XML default' 6"/>
      <action application="bridge" data="loopback/9664"/>
    </condition>
  </extension>
</include>
```

```
<include>
  <extension name="fax detect test">
    <condition field="destination_number" expression="123456">
      <action application="answer"/>
      <action application="set" data="transfer_ringback=${us-ring}"/>
      <action application="spandsp_start_fax_detect" data="transfer '9178 XML de
fault' 6"/>
      <action application="bridge" data="loopback/9664"/>
    </condition>
  </extension>
</include>
<witch/conf/dialplan/default/05_Custom.xml" 10L, 410C written 1,2                All
```

Here, we tell the system to transfer to an extension (9178 for receiving fax) if we detect a fax tone, otherwise the bridge occurs normally (in this case, to a loopback that plays hold music, but you can perform a regular bridge to a phone). You can adapt this principle for use in your own dialplans. Simply create a "fax handler" extension and use `spandsp_start_fax_handler` to transfer to the handler extension whenever a fax machine is detected.

Now the system will automatically handle incoming faxes.

Processing a received fax

Once a fax is received, it rarely needs to just sit somewhere in a directory. Usually, you will want a person to see it transmitted. A common practice is to convert the TIFF file into a PDF file and then email the PDF file as an attachment. Also, users appreciate it when caller ID information can be placed in the subject line of the e-mail. Keep in mind that this will work only if you have a properly configured **mail transport agent MTA**)—for example, `sendmail` or `postfix`—on your system. Create a fax receive extension, as follows:

```
<include>
<extension name="fax_receive">
  <condition field="destination_number"
             expression="^9999$">
    <action application="set"
            data="api_hangup_hook=system
                  ${base_dir}/scripts/emailfax.sh
                  ${fax_remote_station_id}
                  ${base_dir}/faxes/${uuid}.tif"/>
    <action application="playback" data="silence_stream://2000"/>
    <action application="rxfax"
            data="${base_dir}/faxes/${uuid}.tif"/>
    <action application="hangup"/>
  </condition>
</extension>
</include>
```

Note that we've added an `api_hangup_hook` to the fax receive extension. This will cause the `emailfax.sh` script to be executed. Create this script in a text editor and add these lines (you will need to install ghostscript and libtiff-tools for `tiff2pdf`):

```
#!/bin/bash
#
# $1 is the calling fax machine's station ID
# $2 is filename
tiff2pdf -t "Fax from $1" -f -o $2.pdf $2
mutt -n -f /dev/null -F ~/.muttrc -a $2.pdf -s "Fax from $1"
user@domain.com < /dev/null
```

Be sure to replace `user@domain.com` with a valid e-mail address. Finally, create the `.muttrc` file in the home directory and add the following lines:

```
set from = 'sender@domain'
set realname = 'Organization or business name'
set folder = /dev/null
```

Received faxes will now be sent to the specified user with the calling fax machine's station ID.

 Many scripting languages, such as Perl, Python, and Ruby, have libraries that allow you to send e-mails. Feel free to try replacing `emailfax.sh` with your own e-mail sender script.

See also

▶ Refer to the *Sending faxes* recipe in this chapter

Basic text-to-speech with mod_flite

Sometimes, you need a fast, simple, and free text-to-speech implementation for some quick testing. In FreeSWITCH, you can use `mod_flite` for simple TTS testing. While it is not suitable for professional production environments, it meets the criteria of being quick, easy, and free.

Getting ready

Other than a phone and a text editor, there is not much that you need. Keep in mind that on Windows, the `mod_flite` module is prebuilt, but it is not automatically loaded. On Linux/Unix systems, you will need to perform a few steps, as follows.

How to do it...

If you are using Windows, then skip to step 3. If you have Linux/Unix, then follow these steps to enable `mod_flite`:

1. Open `modules.conf` in the FreeSWITCH source, and uncomment the line with `#asr_tts/mod_flite` by removing the # sign at the beginning of the line.

2. Save and exit. Then run the `install` command:

   make mod_flite-install

3. If you wish to have `mod_flite load` by default when FreeSWITCH starts, then open `conf/autoload_configs/modules.conf.xml` and uncomment this line:

```
<!-- <load module="mod_flite"/> -->
```

4. Save and exit. In `fs_cli`, issue the `load mod_flite` command.

At this point, `mod_flite` is active and ready to be used. Now let's add a simple dialplan extension that will let us test it:

1. Open `conf/dialplan/default/066_Custom.xml` and add this extension:

```xml
<include>
<extension name="mod_flite example">
  <condition field="destination_number"
     expression="^(5008)">
    <action application="answer"/>
    <action application="sleep" data="500"/>
    <action application="speak"
         data="flite|kal|Hello world. This is a FreeSWITCH test."/>
  </condition>
</extension>
</include>
```

2. Save the file and exit. Issue the `reloadxml` command from `fs_cli`.

You are now ready to test. Simply dial *5008* and listen to the voice.

How it works...

FreeSWITCH has a `speak` dialplan application that is used to access any installed TTS engine. It accepts pipe-delimited arguments. Note the line we used in the dialplan:

```xml
<action application="speak"
         data="flite|kal|Hello world. This is a FreeSWITCH
test."/>
```

The first argument is the name of the TTS engine, the second argument is the name of the voice for the TTS engine, and the last argument is the actual text to be spoken. The `sleep` app is optional. However, in many cases, it is necessary to pause momentarily after answering a call to allow the media streams to be established.

 Don't confuse the `speak` dialplan application (TTS) with the `say` application! The `say` application is convenient for saying things such as dates, times, numbers, currency, and so on using prerecorded sound prompts.

Flite comes with four voices that you can try out: `awb`, `kal`, `rms`, and `slt`.

See also

▶ Refer to the *Advanced text-to-speech with mod_tts_commandline* recipe in this chapter

Advanced text-to-speech with mod_tts_commandline

Text-to-speech (**TTS**) applications vary in their quality, complexity, and price. However, one thing that most high-end TTS engines have in common is a command-line interface for generating audio from text. FreeSWITCH's mod_tts_commandline module is designed to take advantage of this. While it is completely possible to create a separate module for each engine—and indeed this is the case for mod_flite—it is convenient to utilize a more generic interface that is somewhat agnostic to the exact TTS engine being used.

In this recipe, we will install mod_tts_commandline and then download a free TTS engine that has a command-line interface for use with it. We will also show command-line examples of using some commercial TTS engines.

Getting ready

This recipe has a few prerequisites. The most important one is to get a copy of the freeswitch-contrib git repository. The "contrib repo," as community members call it, contains a number of items given back freely to the FreeSWITCH community as a whole. One of these will assist us with installing the Pico TTS engine, which is part of the Android project. The basic command required to clone the git repository is:

```
git clone https://freeswitch.org/stash/scm/fs/freeswitch-contrib.git
freeswitch-contrib
```

The subdirectory created will simply be referred to as freeswitch-contrib.

How to do it...

If you are using Windows, then skip to step 3. If you have Linux/Unix, then follow these steps to enable mod_tts_commandline:

1. Open modules.conf in the FreeSWITCH source, and uncomment the line with #asr_tts/mod_tts_commandline by removing the # sign at the beginning.

2. Save and exit. Next, run the install command:

   ```
   make mod_tts_commandline-install
   ```

3. If you wish to have `mod_tts_commandline` load by default when FreeSWITCH starts, then open `conf/autoload_configs/modules.conf.xml` and uncomment this line:

```
<!-- <load module="mod_tts_commandline"/> -->
```

4. Save the file and close.

5. Open `conf/autoload_configs/tts_commandline.xml` and locate the line beginning with `<param name="command"...`. Change it to the following:

```
<param name="command" value="pico2wave -w ${file} ${text}
"/>
```

6. For Windows, use `pico2wave.exe` instead of `pico2wave`.

7. Save the file and exit.

At this point, `mod_tts_commandline` is compiled and almost ready for use. Next, let's get the pico TTS engine. Debian/Ubuntu users have to follow these steps:

1. Enable the "non-free" repositories (add `non-free` at the end of each line in `/etc/apt/sources.list`).

2. As root, (or sudo) execute this line:

   apt-get install libttspico-utils

Windows users will need to locate the appropriate solution file in `freeswitch-contrib\grmt\mod_tts_commandline` for Windows:

▶ `mod_tts_commandline.2008.vcproj` for Visual Studio 2008

▶ `mod_tts_commandline.2010.vcxproj` for Visual Studio 2010

Open the appropriate solution file and then rebuild.

You will now have the `pico2wave` (or `pico2wave.exe` in Windows) command-line utility.

Now let's add a simple dialplan extension that will let us use `tts_commandline` and `pico`:

1. Create `conf/dialplan/default/07_Custom.xml`. Add this extension to it:

   ```xml
   <extension name="mod_tts_commandline example">
     <condition field="destination_number"
       expression="^(5010)">
       <action application="answer"/>
       <action application="sleep" data="2000"/>
       <action application="speak"
   data="tts_commandline|pico|Hello
       world. This is a FreeSWITCH test."/>
     </condition>
   </extension>
   ```

2. Save the file and exit. Issue the `reloadxml` command from `fs_cli`.

3. At `fs_cli`, issue the `load mod_tts_commandline` command.

You are now ready to test. Simply dial *5010* and listen to the voice.

How it works...

There are several elements that interact to make this work. We first built `mod_tts_commandline` (just as we would in any other FreeSWITCH module), and then configured it to use `pico2wave` or `pico2wave.exe`. Next, we installed the `pico2wave` command-line utility. Finally, we created a simple dialplan to call the speak application and read our text.

Pico can read text in German, English (GB and US), Spanish, French, and Italian. You can specify in `mod_tts_commandline.conf.xml` which language to use, adding an option like "-l en-GB," as shown here:

```xml
<param name="command" value="pico2wave -l it-IT -w ${file}
${text} "/>
```

There's more...

The really interesting part of `mod_tts_commandline` occurs in the configuration file. The command parameter tells `mod_tts_commandline` what to execute when the `speak` application is called. Read on for some tricks that you can perform with `tts_commandline.conf.xml`.

Modifying the audio stream

It is possible to use an intermediate program, such as **Sound eXchange (SoX)**, to modify the audio that is output from `pico2wave`. An example of this is to resample the audio. By default, `pico2wave` generates mono 16 kHz wave files. If the audio you hear from `mod_tts_commandline` sounds too fast or too slow, then try resampling with SoX. Open `conf/autoload_configs/tts_commandline.conf.xml`, and modify the command parameter.

For Linux/Unix, use this entry:

```
<param name="command" value="pico2wave -w /tmp/$$.wav ${text}
&& sox /tmp/$$.wav -r ${rate} ${file} && rm /tmp/$$.wav"/>
```

For Windows, use the following entry:

```
<param name="command" value="pico2wave.exe -w c:\\tmp\\$$.wav
${text} && sox.exe C:\\tmp\\$$.wav -r ${rate} ${file} && del
c:\\tmp\\$$.wav"/>
```

Ensure that `C:\tmp` exists, or use an appropriate folder on your Windows system.

You will need to issue `reloadxml` at `fs_cli` as well as `reload mod_tts_commandline` for the changes to take effect.

> SoX can perform an amazing array of effects on an audio stream. You can learn more about it at `http://sox.sourceforge.net/`.

Other TTS engines

The FreeSWITCH community has tested `mod_tts_commandline` with a number of commercial TTS engines, mostly under Linux environments. If you have one of the following TTS engines, then use one of the command parameter entries listed in the next code block. In some cases, you will need to tweak your command-line parameters. A simple way to test is to manually run your command and generate a .wav file on the disk, such as `/tmp/test.wav`. Then use a simple dialplan snippet to play back the file:

```
<condition field="destination_number" expression="^(5010)$">
  <action application="answer"/>
  <action application="sleep" data="500"/>
```

```
        <action application="playback" data="/tmp/test.wav"/>
    </condition>
```

This is much easier than making repeated changes to `tts_commandline.conf.xml` and reloading `mod_tts_commandline`. Once you have perfected your command-line syntax, update the configuration file and test.

Configuration file examples

Examples of configuration files are as follows:

> ▸ **Festival**: This is the same engine used in `mod_flite`:
>
> ```
> <param name="command" value="echo ${text} | text2wave -f ${rate} >
> ${file}"/>
> ```
>
> ▸ **Cepstral**:
>
> ```
> <param name="command" value="swift -n ${voice} ${text}
> -o ${file}"/>
> ```
>
> ▸ **Loquendo**:
>
> ```
> <param name="command" value="echo ${text} |
>     TTSFileGenerator -v${voice} -o${file}"/>
> ```

See also

> ▸ Refer to the *Basic text-to-speech with mod_flite* recipe earlier in this chapter

Recording calls

Many enterprises need to record calls for quality control purposes. This recipe describes how you can record inbound and outbound calls on your FreeSWITCH server. If you need assistance in getting calls into and out of your FreeSWITCH system, refer to *Inbound DID (also known as DDI) calls* and *Outgoing calls*, both in *Chapter 1, Routing Calls*.

 Most countries and localities have laws related to recording of phone calls. Always consult a licensed legal professional in your jurisdiction before you start recording phone calls.

Getting ready

Recording calls is actually very simple. All you need is a text editor so that you can add a few lines to your dialplan.

How to do it...

The `record_session` FreeSWITCH dialplan application is used to record calls, whether they are inbound or outbound (the call direction does not affect the `record_session` application).

For inbound calls, it is easiest to enable recording right on `Local_Extension`. Follow the steps:

1. Open `conf/dialplan/default.xml` and locate the `Local_Extension` dialplan entry. Add these lines right before the line with the first bridge application:

   ```
   <action application="set"
   data="record_file_name=$${recordings_dir}/${strftime(%Y-%m-%d-%H-
   %M-%S)}_${uuid}.wav" inline="true"/>
   <action application="record_session"
     data="${record_file_name}"/>
   ```

2. Save the file. Then run `fs_cli` and `r` issue the `reloadxml` command.

Now, any call made to a local extension will be recorded (this includes internal calls from one phone extension to another).

For outbound calls, we need to do something a bit different, because we don't know for sure that the call will actually be answered:

1. Open the dialplan file that contains your outbound route. Add these lines right before your `bridge` application:

   ```
   <action application="set"
      data="record_file_name=$${recordings_dir}/${strftime(%Y-%m-%d-
   %H-%M-%S)}_${uuid}.wav" inline="true"/>
   <action application="export"
        data="execute_on_answer=record_session ${record_file_name}"/>
   ```

2. Save the file, then run `fs_cli`, and issue the `reloadxml` command.

Now, any answered call made through this gateway will be recorded.

How it works...

The `record_session` application will record the audio on the channel. Technically, this application is only running on one leg of the call. In the inbound example, it is running on the called leg (the B leg), but in the outbound example, it is running on the calling leg (the A leg). The `record_session` application records audio in both directions, and therefore, the entire call is recorded.

The filename is stored in the `record_file_name` channel variable. We piece together several bits of information to create the full path:

- `$${recordings_dir}`: By default, this gets set to `$${base_dir}/recordings/`
- `strftime(%Y-%m-%d-%H-%M-%S)`: This produces a timestamp in the format of `YYYY-MM-DD-hh-mm-ss`
- `${uuid}.wav`: This adds the calls' unique ID to the filename

The net result is that our file has a complete and unique path and filename, as follows, this for example:

/usr/local/freeswitch/recordings/2015-05-29-06-54-16_34822476-aefb-4c6a-b1ce-60752ad03768.wav

> The `strftime` API is very handy for getting the current date and time in various formats. It uses the format strings found in the standard `strftime` Unix command. You can experiment with it in `fs_cli`. Try issuing different commands, such as `strftime` and `strftime %Y-%m-%d-%H-%M-%S`, to see what you get.

There's more...

You may have noticed that `Local_Extension` has a curious entry:

```
<action application="bind_meta_app" data="2 b s record_
session::$${recordings_dir}/${caller_id_number}.${strftime(%Y-%m-%d-
%H-%M-%S)}.wav"/>
```

By default, a user who receives a call can manually enable the call recording by pressing *2. By itself, this is a handy feature. However, in the case where we automatically record all calls, this feature is irrelevant. A much more useful feature would be the ability to turn off the call recording. This can easily be done by adding a few more lines to our dialplan. Note that we only want our telephone user (whom we usually call an "agent") to be able to control the call recording. This means that we need to enable a key combination only on the agent's leg of the call. The agent is the A leg on an outbound call and is the B leg on an inbound call. Fortunately, we already have separate dialplan entries for each call type. We simply need to add the appropriate `bind_meta_app` in each case.

For inbound calls, we need to replace the aforementioned `bind_meta_app` entry.
Open `conf/dialplan/default.xml` and replace the `curious` entry with this line:

```
<action application="bind_meta_app" data="2 b s
execute_extension::stop_recording_${dialed_extension} XML
recordings"/>
```

Save the file and exit. For outbound calls, open the dialplan file to which you added the
`record_session` application. Add the following line right before the `bridge` application:

```
<action application="bind_meta_app" data="2 a s execute_
extension::stop_recording_${caller_id_number} XML recordings"/>
```

Save the file and exit. The last step is to create a new dialplan file that will handle the "stop
recording" action that we have implemented. Create a new file in `conf/dialplan/` called
`recordings.xml`, and add these lines:

```
<include>
  <context="recordings">
    <extension name="Stop Recording"/>
      <condition field="destination_number"
      expression="^stop_recording_(.*)$">
        <action application="log" data="WARNING Agent $1 has stopped
        a recording"/>
        <action application="stop_record_session"
        data="${record_filename}"/>
        <action application="set" data="res=${uuid_broadcast
          ${uuid} ivr/ivr-recording_stopped.wav both}"/>
      </condition>
    </extension>
  </context>
</include>
```

Save the file and exit. Then open `fs_cli` and issue the `reloadxml` command. Now test the
feature. Have an agent press *2 on an active call. The agent and caller/callee should hear,
"Recording stopped." The console will show the `stop_record_session` application being
executed. Now the call recording will be stopped.

See also

- Refer to the *Incoming DID (also known as DDI) calls* and the *Outgoing calls* recipes in
  Chapter 1, Routing Calls

6

WebRTC and Mod_Verto

In this chapter, we will cover, the following recipes:

- ► Configuring FreeSWITCH for WebRTC
- ► SIP signaling in JavaScript with SIP.js (WebRTC client)
- ► Verto installation and setup
- ► Verto signaling in JavaScript using Verto.js (Verto client)

Introduction

FreeSWITCH is both a WebRTC gateway and a WebRTC application server. It throws in the signaling plane too, with Verto. Let's introduce these concepts/functions:

- ► FreeSWITCH is a **WebRTC gateway** because it's able to accept encrypted media from browsers, convert it, and exchange it with other communication networks that use different codecs and encryptions, for example, PSTN, mobile carriers, legacy systems, and others. FreeSWITCH can be a gateway between your SIP network and applications and billions of browsers on desktops, tablets, and smartphones.

- ► FreeSWITCH is a **WebRTC application server** because it's able to directly provide native services to browsers, such as video conferences, IVRs, and call centers, without the use of any gateway or third-party. FreeSWITCH can directly provide services through Secure WebSocket (WSS), SRTP, and DTLS, the native WebRTC protocols.

> ▶ FreeSWITCH throws in the **Signaling Plane** because, with Verto, browsers can initiate or receive a voice call or video call in the easiest way, and they can chat, share the screen, and receive and send data in real time to backend applications. Verto is an alternative to XMPP and SIP in JavaScript. FreeSWITCH can serve in parallel and concurrently the same application to clients that use signaling in SIP and Verto.

Audio and Video Communication in the browser

Let's see the basic steps needed to activate WebRTC on FreeSWITCH.

Configuring FreeSWITCH for WebRTC

WebRTC is all about security and encryption. These are not an afterthought. They're intimately interwoven at the design level and are mandatory. For example, you cannot stream audio or video clearly (without encryption) via WebRTC.

Getting ready

To start with this recipe, you need certificates. These are the same kind of certificates used by web servers for SSL-HTTPS.

Yes, you can be your own Certification Authority and self-sign your own certificate. However, this will add considerable hassles; browsers will not recognize the certificate, and you will have to manually instruct them to make a security exception and accept it, or import your own Certification Authority chain to the browser. Also, in some mobile browsers, it is not possible to import self-signed Certification Authorities at all.

The bottom line is that you can buy an SSL certificate for less than $10, and in 5 minutes. (No signatures, papers, faxes, telephone calls... nothing is required. Only a confirmation email and a few bucks are enough.) It will save you much frustration, and you'll be able to cleanly showcase your installation to others.

The same reasoning applies to DNS **Full Qualified Domain Names** (**FQDN**)—certificates belonging to FQDN's. You can put your DNS names in /etc/hosts, or set up an internal DNS server, but this will not work for mobile clients and desktops outside your control. You can register a domain, point an fqdn to your machine's public IP (it can be a Linode, an AWS VM, or whatever), and buy a certificate using that fqdn as **Common Name** (**CN**).

> Don't try to set up the WebRTC server on your internal LAN behind the same NAT that your clients are into (again, it is possible but painful).

How to do it...

Once you have obtained your certificate (be sure to download the Certification Authority Chain too, and keep your Private Key; you'll need it), you must concatenate those three elements to create the needed certificates for mod_sofia to serve SIP signaling via WSS and media via SRTP/DTLS.

```
/usr/local/freeswitch/certs/wss.pem # CERT, KEY AND CHAIN files separated by \n
/usr/local/freeswitch/certs/agent.pem # CERT file AND key file separated by \n
/usr/local/freeswitch/certs/cafile.pem # CHAIN file or root CA
```

With certificates in the right place, you can now activate ssl in Sofia. Open /usr/local/ freeswitch/conf/vars.xml:

```
AMZ                                                          _ □ ×

...
<!-- Internal SIP Profile -->
<X-PRE-PROCESS cmd="set" data="internal_auth_calls=true"/>
<X-PRE-PROCESS cmd="set" data="internal_sip_port=5060"/>
<X-PRE-PROCESS cmd="set" data="internal_tls_port=5061"/>
<X-PRE-PROCESS cmd="set" data="internal_ssl_enable=false"/>

<!-- External SIP Profile -->
<X-PRE-PROCESS cmd="set" data="external_auth_calls=false"/>
<X-PRE-PROCESS cmd="set" data="external_sip_port=5080"/>
<X-PRE-PROCESS cmd="set" data="external_tls_port=5081"/>
<X-PRE-PROCESS cmd="set" data="external_ssl_enable=false"/>
                                              418,2              95%
```

As you can see, in the default configuration, both lines that feature SSL are false. Edit them both to change them to true.

How it works...

By default, Sofia will listen on port 7443 for WSS clients. You may want to change this port if you need your clients to traverse very restrictive firewalls. Edit `/usr/local/freeswitch/conf/sip-profiles/internal.xml` and change the "wss-binding" value to 443. This number, 443, is the HTTPS (SSL) port, and is almost universally open in all firewalls. Also, WSS traffic is indistinguishable from `https/ssl` traffic, so your signaling will pass through the most advanced **Deep Packet Inspection**. Remember that if you use port 443 for WSS, you cannot use that same port for HTTPS, so you will need to deploy your secure web server on another machine.

There's more...

A few examples of such a configuration are certificates, DNS, and STUN/TURN.

Generally speaking, if you set up with real DNS names, you will not need to run your own **STUN** server; your clients can rely on Google STUN servers. But if you need a **TURN** server because some of your clients need a media relay (which is because they're behind and demented NAT got UDP blocked by zealous firewalls), install on another machine `rfc5766-turn-server`, and have it listen on TCP ports 443 and 80. You can also put certificates with it and use TURNS on encrypted connection. The same firewall piercing properties as per signaling.

SIP signaling in JavaScript with SIP.js (WebRTC client)

Let's carry out the most basic interaction with a web browser audio/video through WebRTC. We'll start using `SIP.js`, which uses a protocol very familiar to all those who are old hands at VoIP.

A web page will display a **click-to-call** button, and anyone can click for inquiries. That call will be answered by our company's PBX and routed to our employee extension (1010). Our employee will wait on a browser with the "answer" web page open, and will automatically be connected to the incoming call (if our employee does not answer, the call will go to their voicemail).

Getting ready

To use this example, download version 0.7.0 of the `SIP.js` JavaScript library from `www.sipjs.com`.

We need an "anonymous" user that we can allow into our system without risks, that is, a user that can do only what we have preplanned. Create an anonymous user for click-to-call in a file named `/usr/local/freeswitch/conf/directory/default/anonymous.xml`:

```
<include>
  <user id="anonymous">
    <params>
      <param name="password" value="welcome"/>
    </params>
    <variables>
      <variable name="user_context" value="anonymous"/>
      <variable name="effective_caller_id_name" value="Anonymous"/>
      <variable name="effective_caller_id_number" value="666"/>
      <variable name="outbound_caller_id_name" value="$${outbound_caller_name}"/>
      <variable name="outbound_caller_id_number" value="$${outbound_caller_id}"/>
    </variables>
  </user>
</include>
```

Then add the user's own dialplan to `/usr/local/freeswitch/conf/dialplan/anonymous.xml`:

```
<include>
  <context name="anonymous">
    <extension name="public_extensions">
      <condition field="destination_number" expression="^(10[01][0-9])$">
        <action application="transfer" data="$1 XML default"/>
      </condition>
    </extension>
    <extension name="conferences">
      <condition field="destination_number" expression="^(36\d{2})$">
        <action application="answer"/>
        <action application="conference" data="$1-${domain_name}@video-mcu"/>
      </condition>
    </extension>
    <extension name="echo">
      <condition field="destination_number" expression="^9196$">
        <action application="answer"/>
        <action application="echo"/>
      </condition>
    </extension>
  </context>
</include>
```

How to do it...

In a directory served by your HTPS server (for example, Apache with an SSL certificate), put all the following files.

Minimal click-to-call caller client

HTML (`call.html`):

```html
<html>
<body>
        <button id="startCall">Start Call</button>
        <button id="endCall">End Call</button>
        <br/>
        <video id="remoteVideo"></video>
        <br/>
        <video id="localVideo" muted="muted" width="128px"
height="96px"></video>
        <script src="js/sip-0.7.0.min.js"></script>
        <script src="call.js"></script>
</body>
</html>
```

JAVASCRIPT (`call.js`):

```javascript
var session;

var endButton = document.getElementById('endCall');
endButton.addEventListener("click", function () {
        session.bye();
        alert("Call Ended");
}, false);

var startButton = document.getElementById('startCall');
startButton.addEventListener("click", function () {
        session = userAgent.invite('sip:1010@gmaruzz.org', options);
        alert("Call Started");
}, false);

var userAgent = new SIP.UA({
        uri: 'anonymous@gmaruzz.org',
        wsServers: ['wss://self2.gmaruzz.org:7443'],
        authorizationUser: 'anonymous',
        password: 'welcome'
});
```

```
var options = {
      media: {
            constraints: {
                  audio: true,
                  video: true
            },
            render: {
                  remote: document.
getElementById('remoteVideo'),
                  local: document.getElementById('localVideo')
            }
      }
};
```

Minimal callee

HTML (answer.html):

```
<html>
<body>
      <button id="endCall">End Call</button>
      <br/>
      <video id="remoteVideo"></video>
      <br/>
      <video id="localVideo" muted="muted"
width="128px" height="96px"></video>
      <script src="js/sip-0.7.0.min.js"></script>
      <script src="answer.js"></script>
</body>
</html>
```

JAVASCRIPT (answer.js):

```
var session;

var endButton = document.getElementById('endCall');
endButton.addEventListener("click", function () {
      session.bye();
      alert("Call Ended");
}, false);

var userAgent = new SIP.UA({
      uri: '1010@gmaruzz.org',
      wsServers: ['wss://self2.gmaruzz.org:7443'],
      authorizationUser: '1010',
      password: 'ciaociao'
});
```

```
userAgent.on('invite', function (ciapalo) {
        session = ciapalo;
        session.accept({
             media: {
                     constraints: {
                             audio: true,
                             video: true
                     },
                     render: {
                             remote:
document.getElementById('remoteVideo'),
                             local:
document.getElementById('localVideo')
                     }

             }
        });
});
```

How it works...

Our employee (the callee, or the person who will answer the call) will sit tight with the
`answer.html` web page open on their browser. Upon page load, JavaScript will have created
the SIP agent and registered it with our FreeSWITCH server as SIP user "1010" (just as our
employee was on their own regular SIP phone).

Our customer (the caller, or the person who initiates the communication) will visit the `call.html` webpage (while loading, this web page will register as an SIP "anonymous" user with FreeSWITCH), and then click on the **Start Call** button. This clicking will activate the JavaScript that creates the communication session using the `invite` method of the user agent, passing as an argument the SIP address of our employee.

The **Invite** method initiates a call, and our FreeSWITCH server duly invites SIP user 1010. That happens to be the `answer.html` web page our employee is in front of.

The Invite method sent from FreeSWITCH to `answer.html` will activate the JavaScript local user agent, which will create the session and accept the call.

At this moment, the caller and callee are connected, and voice and video will begin to flow back and forth. The received audio or video stream will be rendered by the `RemoteVideo` tag in the web page, while its own stream (the video that is sent to the peer) will show up locally in the little `localVideo` tag. That's muted not to generate Larsen whistles.

See also

The *Configuring an SIP phone to register with FreeSWITCH* recipe in *Chapter 2, Connecting Telephones and Service Providers*, and the documentation at `http://sipjs.com/guides/`.

Verto installation and setup

Verto is a web protocol that allows the presence of richer distributed applications, where audio, video, database, graphics, visualization, and augmented reality come through the same channel/tool set and converge in the browser for the user to interact with.

Verto can subscribe to data structures residing in the server, and have those structures bi-directionally updated and synchronized in real time.

It is deeply integrated with the FreeSWITCH events system, so it can access all of the data, primitives, statuses, and information that is available for a FreeSWITCH server-side application programmer.

Also, it's very easy to write a FreeSWITCH module that will make **Verto** exchange data structures and events with external sources and sinks, from databases to legacy systems and automation control to collaboration platforms.

Verto is an elegant weapon. It is depicted in the following diagram:

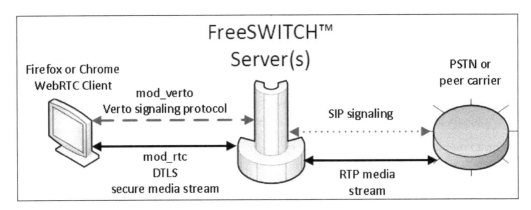

`mod_verto` is compiled by default in the standard installation, but you need to configure it.

In order not to add unwarranted complexities, and to communicate securely through WSS (Secure WebSockets), Verto uses the same SSL certificates as sofia, specifically `/usr/local/freeswitch/certs/wss.pem` (that's a concatenation of all three elements—cert, key, and chain—in the given order).

Here, I will not renew my enumeration of the reasons for favoring a real, bought SSL certificate over a self-signed one (see earlier, in this chapter). I'm a Cheap Charlie, and if I tell you that it is $10 well invested, trust me.

How to do it...

Edit the `/usr/local/freeswitch/conf/autoload_configs/verto.conf.xml` file, as depicted in the following screenshot:

```
AMZ                                                                    _ □ ×
<configuration name="verto.conf" description="HTML5 Verto Endpoint">

  <settings>
    <param name="debug" value="0"/>
  </settings>

  <profiles>
    <profile name="mine">
      <param name="bind-local" value="0.0.0.0:8081"/>
      <param name="bind-local" value="0.0.0.0:8082" secure="true"/>
      <param name="force-register-domain" value="$${domain}"/>
      <param name="secure-combined" value="$${certs_dir}/wss.pem"/>
      <param name="secure-chain" value="$${certs_dir}/wss.pem"/>
      <param name="userauth" value="true"/>
      <param name="ext-rtp-ip" value="52.28.83.83"/>
      <!-- setting this to true will allow anyone to register even with no account
so use with care -->
      <param name="blind-reg" value="false"/>
      <param name="mcast-ip" value="224.1.1.1"/>
      <param name="mcast-port" value="1337"/>
      <param name="rtp-ip" value="$${local_ip_v4}"/>
      <!-- <param name="ext-rtp-ip" value=""/> -->
      <param name="local-network" value="localnet.auto"/>
      <param name="outbound-codec-string" value="pcmu,opus,vp8"/>
      <param name="inbound-codec-string" value="pcmu,opus,vp8"/>
      <param name="apply-candidate-acl" value="wan.auto"/>
      <param name="timer-name" value="soft"/>

    </profile>
  </profiles>

</configuration>
<switch/conf/autoload_configs/verto.conf.xml" 31L, 1286C written 3,3        All
```

How it works...

The most important parameters here are as follows:

- **debug**: Set this to "10" if you need to follow the inner workings and see what the JSONs exchanged back and forth with the browsers. From registration to call initiation, and the "floor" status in conferences, it's all here.

- **bind-local, "secure=true"**: This is the port used for WSS listening.

- **ext-rtp-ip**: This is the public IP address of FreeSWITCH. In a NATted, installation is *not* the IP address of the physical network interface, but the external address that the router or firewall then redirects here. You need to set this correctly (for example, in Amazon AWS or in your data center LAN).

▶ **outbound and inbound codec strings**: These define which codecs are negotiated (directly or via ICE) before exchanging media. In native WebRTC, there is a marked preference for the OPUS codec, but G711 (PCMU) is the lingua franca of interoperability with all communication worlds.

As always, after editing the configuration file, reload the module ("reload `mod_verto`" from `fs_cli`) or restart FreeSWITCH.

Verto signaling in JavaScript using Verto.js (Verto client)

Let's implement a click-to-call button that will allow anyone to click on a web page to join a conference in video-audio chat (more information about conferences is covered in the next chapter).

For security reasons, we need an "anonymous" user that we can allow into our system without risks, that is, a user that can do only what we have preplanned.

Create an anonymous user for click-to-call by adding to the directory and the dialplan as instructed in the *SIP Signaling in JavaScript with SIP.js(WebRTC client)* section. Then use `reloadxml` from `fs_cli`, or restart FreeSWITCH.

The end user does not need to log in or identify themselves; **insert your name** is there just for aesthetic purposes.

A minimal click-to-call Verto client looks like this:

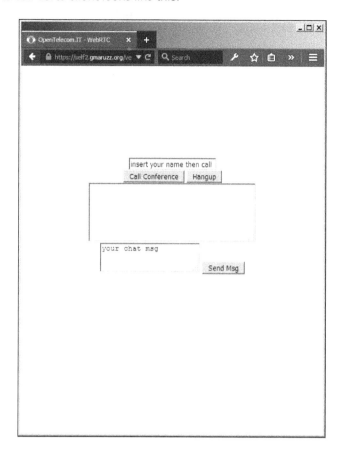

How to do it...

In a directory served by your HTPS server (for example, Apache with an SSL certificate), put these two files:

HTML (3.html):

```
<!DOCTYPE html>
<html>
 <head>
  <meta name="viewport" content="width=device-width, initial-scale=1,
user-scalable=yes"/>
  <link rel="shortcut icon" href="favicon.ico" />
  <title>OpenTelecom.IT - WebRTC</title>
 </head>
 <body>
```

```html
    <center>
     <input type="hidden" id="hostname" value="self2.gmaruzz.org"/>
     <input type="hidden" id="wsURL" value="wss://self2.gmaruzz.
org:8082"/>
     <input type="hidden" id="login" value="anonymous"/>
     <input type="hidden" id="passwd" value="welcome"/>
     <input type="hidden" id="cidnumber" value="WebRTC"/>
     <input type="hidden" id="ext" value="3600"/>
     <video id="webcam" autoplay="autoplay" style="width:auto;height:aut
o;max-height:100%;max-width:100%;"></video>
     <br/>
     <input type="text" size="22" id="cidname" value="insert your name
then call"/>
     <br/>
     <button data-inline="true"id="callbtn">Call Conference</button>
     <button data-inline="true" id="hupbtn">Hangup</button>
     <br/>
     <textarea id="chatwin" style="width:300px;height:100px;scrolling=au
to;"></textarea>
     <br/>
     <textarea id="chatmsg">your chat msg</textarea>
     <button id="chatsend">Send Msg</button>
    </center>
     <script type="text/javascript" src="js/jquery-2.1.1.min.js"></
script>
     <script type="text/javascript" src="js/jquery.json-2.4.min.js"></
script>
     <script type="text/javascript" src="js/jquery.cookie.js"></script>
     <script type="text/javascript" src="js/verto-min.js"></script>
     <script type="text/javascript" src="3.js"></script>
    </body>
    </html>
```

JAVASCRIPT (3.js):

```javascript
    'use strict';
    var cur_call = null;
    var verto;
    var chatting_with = false;

    var callbacks = {
      onMessage: function(verto, dialog, msg, data) {
         console.error("msg ", msg);
         console.error("data ", data);

         switch (msg) {
```

```
      case $.verto.enum.message.pvtEvent:
       if (data.pvtData) {
        console.error("data.pvtData ", data.pvtData);
        switch (data.pvtData.action) {
         case "conference-liveArray-join":
          chatting_with = data.pvtData.chatID;
          break;
        }
       }
       break;
      case $.verto.enum.message.info:
       var body = data.body;
       var from = data.from_msg_name || data.from;

       if (body.slice(-1) !== "\n") {
        body += "\n";
       }
       $('#chatwin')
        .append(from + ': ')
        .append(body)
        $('#chatwin').animate({"scrollTop": $('#chatwin')[0].
scrollHeight}, "fast");
       break;
      default:
       break;
     }
    },
  onEvent: function(v, e) {
   console.error("GOT EVENT", e);
  },
};

function docall() {
 if (cur_call) {
  return;
 }
 cur_call = verto.newCall({
  destination_number: $("#ext").val(),
  caller_id_name: $("#cidname").val(),
  caller_id_number: $("#cidnumber").val(),
  useVideo: true,
  useStereo: true,
  useCamera: $("#usecamera").find(":selected").val(),
  useMic: $("#usemic").find(":selected").val()
 });
}

$("#callbtn").click(function() {
  docall();
```

```
    });

    $("#hupbtn").click(function() {
      verto.hangup();
      cur_call = null;
    });

    function setupChat() {
     $("#chatwin").html("");

     $("#chatsend").click(function() {
       if (!cur_call && chatting_with) {
       return;
     }
       cur_call.message({to: chatting_with,
         body: $("#chatmsg").val(),
         from_msg_name: cur_call.params.caller_id_name,
         from_msg_number: cur_call.params.caller_id_number
         });
       $("#chatmsg").val("");
     });

     $("#chatmsg").keyup(function (event) {
       if (event.keyCode == 13 && !event.shiftKey) {
       $( "#chatsend" ).trigger( "click" );
       }
     });
    }

    function init() {
     cur_call = null;

     verto = new $.verto({
     login: $("#login").val() + "@" + $("#hostName").val(),
     passwd: $("#passwd").val(),
     socketUrl: $("#wsURL").val(),
     tag: "webcam",
     iceServers: true
     },callbacks);

     $(document).keypress(function(event) {
      var key = String.fromCharCode(event.keyCode || event.charCode);
      var i = parseInt(key);

      if (key === "#" || key === "*" || key === "0" || (i > 0 && i <= 9))
    {
        cur_call.dtmf(key);
      }
     });
```

```
    setupChat();
  }

  $(window).load(function() {
    $.verto.init({}, init);
  });
```

How it works...

HTML is not complex. Let's start from beginning; the `viewport` line makes the page adaptable to both desktops and smartphones. We set some variables needed to log in as an "anonymous" user on our FreeSWITCH WebRTC server. We define the extension to be called in `ext`.

Then, the `video` tag will be the audio/video renderer of the stream we get from the server. The important values are those of `id` (referred to by the Verto constructor in `3.js`), `autoplay` (you would want it to render the stream as soon as possible), and `style` (a carefully chosen size for adaptation to both desktops and phones).

There's input text for the nickname and a couple of buttons: one to start the call, and the other to hang up.

Then, there is a `chat zone` made by a textarea of suitable size to show the conversation, an input text for the message, and a button for sending it.

At the end, the JavaScript files we want to include are referenced.

The JavaScript file is longer than the HTML file, but I tried to keep it straightforward, yet functional. It will deliver a lot of useful output in the browser's JavaScript console.

Let's comment on it, starting from the end and going backward.

When the web page has finished loading, a `verto` object will be created and initialized from the variables we set in HTML. `iceServers` is `true`, so we'll use the default STUN server (we could have passed an array of STUN and TURN servers instead). The last argument for the initialization of the `verto` object is the `callbacks` variable. It is a structure of functions that the verto object executes to react to events. We will look at it in detail later. Then, the event function will send DTMFs to our call when *0-9*, ***, and *#* are pressed (if they're not input in the chat area). Note the compatibility of Firefox and Chrome.

The last line of the verto object initialization calls the setup of the chat-related variables and functions. Particularly, it will trigger the `message()` method of the current call, both when the button is pressed and when *Enter* is pressed in the chat input text area. The `message()` function will use variables set by the `callbacks` functions we gave to the verto object when we created it.

Then, the two functions are executed when the **Call Conference** and **Hangup** buttons are pressed. The `docall` function gets the variables we set in HTML as arguments, and we hardcode some more. `useCamera` and `useMic` are set to the device the user has chosen when asked to give authorization by the browser.

Now, let's cover "callbacks". These are the functions that get automatically executed in real time at messages and events receiving. We are reacting only to `conference-liveArray-join` to set the data structure we're synchronizing with, and to `message.info` in order to display the chat conversation's content. You can see in the browser's JavaScript console some other events and messages that are available for interaction.

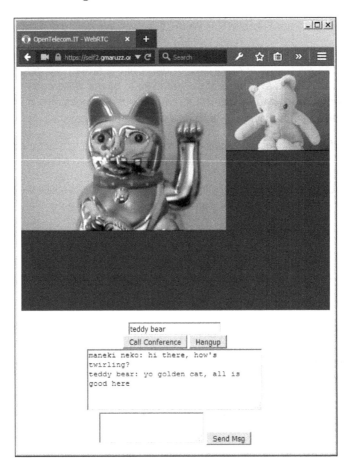

See also

Verto is documented on the Confluence page at `http://freeswitch.org/confluence/display/FREESWITCH/mod_verto`

7

Dialplan Scripting with Lua

In this chapter, we will cover these recipes:

- ▸ Creating a basic Lua script
- ▸ User interaction and DTFM gathering
- ▸ Using conditionals, loops, web calls, and regular expressions
- ▸ Connecting to an external database

Introduction

An XML dialplan is the standard and most efficient way to describe how a call must be handled by FreeSWITCH. There are cases where you need loops, conditionals, and other call handling logic that is not easily expressed in XML. Enter scripting, and you can use your programming language of choice.

FreeSWITCH supports many scripting languages (almost all of them). These languages also allow you to use the same primitives and access the same variables, so they're all functionally equivalent. The difference lies in their efficiency and their ability to embed, for example, how much CPU they consume, and how much RAM they need to execute the same call logic.

FreeSWITCH's most efficient and embeddable scripting language is Lua (www.lua.org).

It is a very easy procedural language, reminiscent of C and Perl, so it will be immediately familiar to most of you.

Creating a basic Lua script

In this recipe, we will be creating a script that uses the basic Lua functionalities.

Getting ready

You don't need to install or configure anything to be ready to execute Lua scripts to handle your calls. Lua is the standard and preferred FreeSWITCH scripting language, which is compiled and embedded by default.

The script will be called by a dialplan extension, and when the script is finished, the call will be automatically hung up (see the *Using conditionals, loops, web calls, and regular expressions* recipe later to learn how to change this behavior and have the dialplan continue handling after the Lua script ends).

Create a dialplan extension "12345," which will call our first Lua script. Edit the `/usr/local/freeswitch/conf/dialplan/default/01_basic_lua.xml` file:

```xml
<include>
  <extension name="Basic Lua Script">
    <condition field="destination_number" expression="^(12345)$">
        <action application="lua" data="basic.lua"/>
    </condition>
  </extension>
</include>
```

Save the file and then, from `fs_cli`, `reloadxml`.

How to do it...

Edit the `/usr/local/freeswitch/scripts/basic.lua` file like this:

```
--Answer the call
session:answer()

--Wait 2 seconds
session:sleep(2000)

--Play a message to caller
session:execute("playback", "ivr/ivr-welcome_to_freeswitch.wav");

--End of script, will automatically hangup
```

`"/usr/local/freeswitch/scripts/basic.lua" 11L, 212C written   1,1              All`

You don't have to reload anything because, in FreeSWITCH, scripts are read from the filesystem and then executed each time anew. So, if you happen to make a mistake or want to change something in the script, just edit it. Your changes will be picked up the next time the script is executed.

Also, you don't need to make the script "executable." It's just text that will be read and executed by FreeSWITCH and has got a Lua interpreter embedded in it at compile time.

How it works...

The following sequence shows the working of our recipe.

1. When an incoming call hits FreeSWITCH, it will arrive in the dialplan, looking for an extension matching its destination number.

   If the call's destination number is 12345, it will enter our newly created extension that will execute the application lua (the embedded Lua interpreter) with data from basic.lua.

2. FreeSWITCH will then look in the default scripts directory (/usr/local/freeswitch/scripts/) for a basic.lua file to pass to the Lua interpreter.

3. Once the script file is found, Lua interpreter will begin executing it.

   The first script line, session:answer(), will use the answer method of the session object. This object is automatically given to the script by FreeSWITCH, and represents the call leg.

4. The script will then make FreeSWITCH answer the call, for example, go off the hook.

5. Then FS will sleep for 2000 milliseconds (2 seconds), which is good practice, so the audio flow can stabilize between the caller and callee.

6. Then the execute method will be called. It will start an application, exactly as dialplan would have done, passing it arguments (the arguments would be data in dialplan). So, in our case, the playback application is called, with the ivr/ivr-welcome_to_freeswitch.wav argument.

7. It will play the sound file to the caller.

8. After playing the sound file, the script will have finished its instructions, so it will exit.

9. When the script is finished, the call will be automatically hung (see later in the *Using conditionals, loops, web calls, and regular expressions* recipe to learn how to change this behavior, and have dialplan continue handling after the Lua script has ended).

The `session` object, our call leg, is not the only object given for free by FreeSWITCH.

The other one is the `freeswitch` object, which is always present—even if no call leg exists (for example, the script is executed by the FreeSWITCH administrator from the command line, not from the dialplan)—and it gives us access to various facilities, notably to the equivalent of `fs_cli`. From this, we can send commands to FS via the API method.

See also

Refer to *Introduction* in *Chapter 1, Routing Calls,* for a basic understanding of the XML dialplan.

User interaction and DTMF gathering

This recipe elaborates the interaction of the dialplan with the users.

Getting ready

This time, the dialplan extension that we add will do things before starting the Lua script. This is not necessary at all, but is being done here to let you know that the dialplan and scripting complement each other.

> The best strategy for efficiency, counseled by FreeSWITCH developers, is to use the dialplan as much as you can. It is the most efficient (compiled in a tree in RAM) way to process your calls. Then, when things require logic (loops, conditionals, and so on) that does not fit well in XML, spawn a script.

Create a dialplan extension `12346`. It will call our second Lua script. Edit the `/usr/local/freeswitch/conf/dialplan/default/02_interaction.xml` file:

```
<include>
  <extension name="Interaction and DTMFs gathering">
    <condition field="destination_number" expression="^(12346)$">
        <action application="answer"/>
        <action application="sleep" data="2000"/>
        <action application="playback"
data="ivr/ivr-welcome_to_freeswitch.wav"/>
        <action application="set" data="my_channel_variable='foo
bar baz'"/>
```

```
        <action application="lua" data="interaction.lua"/>
      </condition>
    </extension>
  </include>
```

Save the file and then, from `fs_cli`, `reloadxml`.

How to do it...

Edit the `/usr/local/freeswitch/scripts/interaction.lua` file:

```lua
--Wait half seconds
session:sleep(500)

--Play a message to caller
session:execute("playback", "ivr/ivr-douche_telecom.wav");

--Wait half seconds
session:sleep(500)

--get the value set by dialplan into channel
variable_my_channel_variable = session:getVariable("my_channel_
variable");

--print that value on FreeSWITCH console and logfile
session:consoleLog("info", "dialplan set channel variable value to: "
.. variable_my_channel_variable .. "\n");

--prompt the user to enter digits, manage input errors, transfer on
final failure
digits = session:playAndGetDigits(4, 10, 3, 3000, "#",
"ivr/ivr-please_enter_the_number_where_we_can_reach_you.wav",
"ivr/ivr-invalid_extension_try_again.wav", "\\d+", "digits_received",
1000,
"5000 XML default");

--print gathered digits on FreeSWITCH console and logfile
session:consoleLog("info", "Lua variable digits is: ".. digits .."\n")

--Wait half seconds
session:sleep(500)

--Play goodbye to caller
session:execute("playback", "ivr/ivr-thank_you.wav");

--End of script, will automatically hangup
```

How it works...

The preceding recipe works in the following sequence:

1. We start with the dialplan, since the incoming call will go there first.

2. The extension we just created will match an incoming call for destination number `12346`, answer the call, wait for 2 seconds, and then play back a message.

3. When the playback is completed, the `set` application will create a channel variable with the name `my_channel_variable`, and assign it the value `foo bar baz`.

4. That newly created channel variable will be part of your call, like all other channel variables, and you can manipulate and use it; for example, it can be a part of custom CDRs (See *Chapter 3, Processing Call Detail Records*).

5. Then, the `lua` application will execute the embedded Lua interpreter, passing `interaction.lua` as the file containing the script.

Here, the dialplan extension has finished its steps, and the ball (actually the call) is passed to the Lua script.

Let's look into `interaction.lua` and follow it line by line:

1. The first script line, `session:sleep(500)`, will use the `sleep` method of the `session` object.

2. The `session` object is automatically given to the script by FreeSWITCH, and represents the call leg.

3. The `sleep` method will wait for half a second, so there will be a pause between the message played by the dialplan and the message that will be played in the next script line. If we don't put this to `sleep`, the two messages here will be played continuously as if they were the same phrase, and that's not what we want.

4. Then, the `execute` method of the `session` object will activate the playback dialplan application, which will get `ivr/ivr-douche_telecom.wav` looking in the standard `/usr/local/freeswitch/sounds/en/us/callie/` directory, choosing by the audio sampling frequency requested. For example, if the incoming call is from PSTN, it will use an 8,000 Hz audio sampling frequency (codec g711, or g729, and so on), and FreeSWITCH will automatically select the `/usr/local/freeswitch/sounds/en/us/callie/ivr/8000/ivr-douche_telecom.wav` audio file.

5. The next line will be another pause.

6. Then, we see how to use `session:getVariable` to access the channel variables. The value of `my_channel_variable`, which was set by the dialplan, will be assigned to a Lua variable named `variable_my_channel_variable`. Now we can use it in our script.

7. And here we are, using it to construct one of the arguments to `session:consoleLog`.

The `consoleLog` method is used to print both in the `fs_cli` console and in FreeSWITCH's log file. This method gets two arguments. The first one is the log level, and the second is the `message` that will belong to the log level. So, in our case, we choose the `info` log level. We construct the second argument to `consoleLog`, the `message`, as a string concatenation of `Lua variable digits is:`, the value of the `variable_my_channel_variable` Lua variable, and a final carriage return.

 Note that `..` is a Lua string concatenation operator, it will join two strings into one.

8. Then comes one of the workhorses of Lua FreeSWITCH's scripting: `playAndGetDigits`. We have implemented it here in its full glory, with all the optional arguments, albeit you will often be omitting some of them. Let's reproduce the entire line here, because it would be impractical to follow the explanation going back and forth over this book's pages:

```
digits = session:playAndGetDigits(4, 10, 3, 3000, "#",
"ivr/ivr-please_enter_the_number_where_we_can_reach_you.wav",
"ivr/ivr-invalid_extension_try_again.wav", "\\d+",
"digits_received", 1000,
"5000 XML default");
```

Let's comment on this line from left to right:

- **digits**: This is the Lua variable that will get (as a string) the DTMFs sent by the user.
- **4**: This is the minimum number of digits we'll accept.
- **10**: This is the maximum number of digits we'll accept (for example, upon receiving the tenth DTMF in this case, we call it a day).
- **3**: This is the number of times we'll try to get the digits from the user before declaring the final failure.
- **3000**: This is the number of milliseconds we wait for a digit (for example, timeout).
- **#**: This is the digits *terminator*. The user will send it to signal that they have finished entering DTMFs (this is useful for entering digits less than the maximum number of digits possible, without waiting for timeout; for example, you can enter `1234#`, and it will immediately be accepted).
- Then there is the **audio** file that will be played as a prompt for the user.
- Then you will find the **audio** file that will be played to the user if the input entered is invalid.

- ❑ **"\\d+"**: This is a regular expression that defines what we consider a "valid input." In this case, all whatever digit in whatever quantity (note that the backslash needs to be escaped by another backslash).

- ❑ **digits_received**: This is a channel variable that will be created and set to the digits string in the case of success, or to `nil` (the Lua null value) in the case of a failure or invalid input. This argument is optional.

- ❑ **1000**: This is the inter-digit timeout. Use it if you want to change the timeout after the first digit has been entered. This argument is optional, and defaults to the main timeout (which will be 3,000 milliseconds in our case).

- ❑ The last optional argument is where the call leg is transferred in the event of a failure. The format is like this: **extension kindofdialplan context**. The default is no transfer; just continue.

Whoah, `playAndGetDigits` is powerful, isn't it?

9. In the next line, we again use `consoleLog` to print the contents of the Lua digits variable.

10. Then comes a pause.

11. Finally we say *Thank you* to our user and (because the script is finished) hang up.

This screenshot shows what an incoming call that inputs `3456#` will look like in `fs_cli`:

```
freeswitch@internal>
freeswitch@internal> fsctl loglevel info
+OK log level: INFO [6]

2015-07-03 12:24:17.649994 [NOTICE] switch_channel.c:1089 New Channel sofia/inte
rnal/1002@192.168.1.125 [de40f31c-ca06-4b93-b9ed-6dacb81b3817]
2015-07-03 12:24:17.709994 [INFO] mod_dialplan_xml.c:642 Processing giovanni <10
02>->12346 in context default
2015-07-03 12:24:17.709994 [NOTICE] sofia_media.c:92 Pre-Answer sofia/internal/1
002@192.168.1.125!
2015-07-03 12:24:17.709994 [NOTICE] mod_dptools.c:1292 Channel [sofia/internal/1
002@192.168.1.125] has been answered
2015-07-03 12:24:33.989993 [INFO] switch_cpp.cpp:1256 dialplan set channel varia
ble value to: 'foo bar baz'
2015-07-03 12:24:39.689946 [INFO] switch_cpp.cpp:1256 Lua variable digits is: 34
56
2015-07-03 12:24:40.729994 [NOTICE] switch_core_state_machine.c:315 sofia/intern
al/1002@192.168.1.125 has executed the last dialplan instruction, hanging up.
2015-07-03 12:24:40.729994 [NOTICE] switch_core_state_machine.c:317 Hangup sofia
/internal/1002@192.168.1.125 [CS_EXECUTE] [NORMAL_CLEARING]
2015-07-03 12:24:40.729994 [NOTICE] switch_core_session.c:1657 Session 22 (sofia
/internal/1002@192.168.1.125) Ended
2015-07-03 12:24:40.729994 [NOTICE] switch_core_session.c:1661 Close Channel sof
ia/internal/1002@192.168.1.125 [CS_DESTROY]
freeswitch@internal>
```

There's more...

The FreeSWITCH logging system has the concept of loglevels. Each log message is pertaining to a loglevel, and will be printed and/or displayed only if the facility (log file, console, and so on) has been set to accept that level (or lower levels). Log levels proceed from debug (which is of the lowest importance) to info, notice, warning, err, crit, and alert, alert being the level of maximum importance. For example, if we set our console to the notice log level, it will display the notice, warning, err, crit, and alert messages, but will not display the debug and info messages.

Using conditionals, loops, web calls, and regular expressions

This recipe helps us demonstrate the loops and other conditionals prevailing in Lua scripts. You will get also familiar with regular expressions and web calls.

Getting ready

Create a dialplan extension 12347 that will call our third Lua script. Edit the /usr/local/freeswitch/conf/dialplan/default/03_advanced.xml file:

```
<include>
  <extension name="Advanced Lua Script">
    <condition field="destination_number" expression="^(12347)$">
        <action application="answer"/>
        <action application="sleep" data="1000"/>
        <action application="playback" data="ivr/ivr-welcome_to_
freeswitch.wav"/>
        <action application="sleep" data="500"/>
        <action application="lua" data="advanced.lua"/>
        <action application="playback" data="ivr/ivr-thank_you.wav"/>
        <action application="hangup"/>
    </condition>
  </extension>
</include>
```

Save the file and then, from fs_cli, reloadxml.

How to do it...

Edit the `/usr/local/freeswitch/scripts/advanced.lua` file:

```lua
-- callback function, see session:setInputCallback below
function key_press(session, input_type, data, args)
        -- if we got DTMF
        if input_type == "dtmf" then
                -- print to console and logfile which key was pressed
                session:consoleLog("warning", "Key pressed: " ..
data["digit"])
                -- set the "finished" variable to true
                finished = true
                -- returning "break" cause the interruption of media
playing
                return "break"
        end
end


---------------------------------------------------------
-- set a function to be called on DTMFs
session:setInputCallback("key_press", "")
-- after script finish, go back to dialplan
session:setAutoHangup(false)
-- create the api object, to be used to emit fs_cli commands
api = freeswitch.API()

-- value we compare against the current price of Google stock
benchmark = 522
-- flag to break from playing and then from "while" loop
finished = false
---------------------------------------------------------


-- play a message
session:execute("playback", "ivr/ivr-when_finished_press_any_key.wav")

-- print message to console and logfile
session:consoleLog("warning", "BEFORE WHILE\n")

-- do until user hangup or variable "finished" is true
while( session:ready() == true and finished == false) do
        -- do an http query and store the result in "google_values"
Lua variable
```

```
        google_values=api:execute("curl", "http://download.finance.
yahoo.com/d/quotes.csv?s=GOOG&f=sl1po")
        -- print variable to console and logfile
        session:consoleLog("warning", "google_values is: " .. google_
values .. "\n")
        -- parse variable into components
        _,_,var1,var2,var3,var4 = string.find(google_
values,"(.-),(.-),(.-),(.*)")
        -- print components to console and logfile
        session:consoleLog("warning", "var1="..var1.." var2="..var2.."
var3="..var3.." var4="..var4.."\n")
        -- "pronounce" component to user
        session:say(var2, "en", "currency", "pronounced")
        -- check if exit time
        if(finished==true)then break end

        -- evaluate if component is less than benchmark
        if(tonumber(var2) < benchmark) then
                -- if component is less than benchmark
                -- play a message
                session:execute("playback", "ivr/ivr-less_than.wav")
                -- check if exit time
                if(finished==true)then break end
        else
                -- if component is equal or more than benchmark
                -- play a message
                session:execute("playback", "ivr/ivr-more_than.wav")
                -- check if exit time
                if(finished==true)then break end
        end

        -- "pronounce" benchmark to user
        session:say(benchmark, "en", "currency", "pronounced")
        -- check if exit time
        if(finished==true)then break end
        -- pause half a second
        session:sleep(500)
end

-- print message to console and logfile
session:consoleLog("warning", "AFTER WHILE\n")

-- pause half seconds
session:sleep(500)

--End of script, we set autohangup to false, will go back to dialplan
```

How it works...

The extension we just created and loaded into the dialplan will match the destination number `12347`, answer the incoming call, play an audio file to the caller, and then pass the call to our new `advanced.lua` script.

An interesting thing here is that the dialplan execution will continue even after the script is finished. This is because in the script, we have executed the `setAutoHangup` method with a `false` argument.

So, after the script exits, the dialplan will play another audio file to the caller, and only then hang up the call.

Let's look at the script now, from the beginning:

1. First of all, we declare a *function* that will be used immediately below, where we set it as a callback to be executed each time there is input.

   The `key_press` function will check what has been thrown at it, and if a DTMF has been received, it will first print on the console and log file the digit that was received. Then it will assign the `true` value to the finished variable. If DTMF was received, the `key_press` function will exit, returning a `break` value.

2. An input callback function returning a `break` value will interrupt the playing of media. For example, if the session (the channel or the call leg; your mental representation can vary) is playing something to the caller (an audio file, music on hold, a tone stream, and so on) and the caller pressing a key activates the callback function that exits with `break`, then the playing will interrupt, and we'll proceed further to the next step in the script.

3. Then we call the `session:setAutoHangup` method with a `false` argument. This will change the default behavior of hanging up the call when the script has finished, and will instead return control to the dialplan.

4. We create an `api` object, which we'll use later, to give commands to FreeSWITCH as we did in `fs_cli`, and get back the result.

5. We set a couple of Lua variables: `benchmark`, which we'll use later to compare the current value of the Google stocks, and `finished`, which we'll use as a flag to know whether the caller has pressed a key (which means that they want to exit the loop).

6. Play `ivr/ivr-when_finished_press_any_key.wav` to the caller, and then log we're about to enter the `while` loop.

7. The `while` loop will run until the `finished` variable is no longer `false`, or until the result of the `session:ready()` method is no longer `true`.

   The `ready()` method of the session can always be called, and will yield a value of `true` if the call leg is connected (that is, it has been answered and has not yet been hung up).

You must always check the value of `session:ready()` when you enter a loop, because if you don't check and the caller hangs up, the script will run forever.

8. We then have this line:

```
google_values=api:execute("curl",
"http://download.finance.yahoo.com/d/quotes.csv?s=GOOG&f=sl1po")
```

Here, we use the `execute` method of the `api` object we created before to send a command with its argument to FreeSWITCH (as we do in `fs_cli`). We put the string returned by FreeSWITCH into the `google_values` variable.

Then we duly print this variable.

9. Next, we want to parse this very `google_values` variable with a regular expression. We'll do it in the following line:

```
_,_,var1,var2,var3,var4 = string.find(google_
values,"(.-),(.-),(.-),(.*)")
```

The `string.find` function is a Lua function that returns the position of the first and last occurrence of a Lua regular expression in a string. Note that Lua regular expressions are different from POSIX and Perl regexes.

Our regular expression will use parentheses to "capture" four values. The first three values will be composed of the minimum number of whatever characters are present until a comma is found. The fourth captured value will stretch from the third comma to the end of the string.

So, in this case, `string.find` will return six values: the two indexes (we will discard them by putting them into the _ Lua dummy variables), and the four captured substrings that we will assign to the `var1..var4` Lua variables.

10. We'll print those variables' contents.

11. We will then read the current Google stock price to the caller:

```
session:say(var2, "en", "currency", "pronounced")
```

The `say` method is used to have FreeSWITCH read values with some "artificial intelligence." The first argument is the string to be read, the language that has to be read, then which kind of string it is (for example, time, date, currency, e-mail address, persons, name spelled, and so on), and then comes the **reading method** (pronounced, iterated, counted, and so on). `Say` is a beautiful method, fully adapted to different languages, with plurals, genders, and so on. Check it out on FreeSWITCH's Confluence page.

12. After this line, we check whether the caller has just pressed a key, interrupting the variable reading and setting the `finished` variable to `true`. If that is what has happened, we break out from the Lua `while` loop.

13. Then we check whether the numeric value of the current stock price string is lower than the `benchmark` numeric variable.

14. Depending on the conditional result, we read a `less than` or `more than` audio file to the caller.

15. We then `say` the benchmark to the caller, implicitly converting its value into a string.

16. Then comes a pause, and the `while` loop is ready to begin again.

17. When we break out from the `while` loop, we print a line to the console, then pause for half a second, and get back control of the call leg to the dialplan.

This is what a call to "12347" will look like in the `fs_cli` console when the caller listens to the loop for two times and half and then presses 5:

```
freeswitch@internal>
2015-07-04 10:59:12.027820 [NOTICE] switch_channel.c:1089 New Channel sofia/inte
rnal/1002@192.168.1.125 [6317b883-48a6-4d4b-aad3-6a31eac88f12]
2015-07-04 10:59:12.067765 [INFO] mod_dialplan_xml.c:642 Processing giovanni <10
02>->12347 in context default
2015-07-04 10:59:12.067765 [NOTICE] sofia_media.c:92 Pre-Answer sofia/internal/1
002@192.168.1.125!
2015-07-04 10:59:12.067765 [NOTICE] mod_dptools.c:1292 Channel [sofia/internal/1
002@192.168.1.125] has been answered
2015-07-04 10:59:21.887819 [WARNING] switch_cpp.cpp:1256 BEFORE WHILE
2015-07-04 10:59:22.107766 [WARNING] switch_cpp.cpp:1256 google_values is: "GOOG
",523.40,521.84,521.08

2015-07-04 10:59:22.107766 [WARNING] switch_cpp.cpp:1256 var1="GOOG" var2=523.40
 var3=521.84 var4=521.08

2015-07-04 10:59:30.047759 [WARNING] switch_cpp.cpp:1256 google_values is: "GOOG
",523.40,521.84,521.08

2015-07-04 10:59:30.047759 [WARNING] switch_cpp.cpp:1256 var1="GOOG" var2=523.40
 var3=521.84 var4=521.08

2015-07-04 10:59:37.927820 [WARNING] switch_cpp.cpp:1256 google_values is: "GOOG
",523.40,521.84,521.08

2015-07-04 10:59:37.927820 [WARNING] switch_cpp.cpp:1256 var1="GOOG" var2=523.40
 var3=521.84 var4=521.08

2015-07-04 10:59:38.727771 [WARNING] switch_cpp.cpp:1256 Key pressed: 5
2015-07-04 10:59:38.727771 [WARNING] switch_cpp.cpp:1256 AFTER WHILE
2015-07-04 10:59:39.767830 [NOTICE] mod_dptools.c:1266 Hangup sofia/internal/100
2@192.168.1.125 [CS_EXECUTE] [NORMAL_CLEARING]
2015-07-04 10:59:39.767830 [NOTICE] switch_core_session.c:1657 Session 22 (sofia
/internal/1002@192.168.1.125) Ended
2015-07-04 10:59:39.767830 [NOTICE] switch_core_session.c:1661 Close Channel sof
ia/internal/1002@192.168.1.125 [CS_DESTROY]
freeswitch@internal>
```

There's more...

Lua's regular expressions are less powerful than Perl's regexes. If you need the full power of PCRE, you can use this construct from the Lua script:

```
session:execute("set", "chan_variable=${regex(" .. my_variable
.. "|^([0-9]{10})$)}")
```

This will create a channel variable named `chan_variable`, and assign to it the result of the Perl compatible regex applied to the `my_variable` Lua variable.

You will then have to access this newly created channel variable:

```
variable_my_channel_variable =
session:getVariable("chan_variable");
```

See also

Refer to the previous recipe to see how to set and access channel variables.

Connecting to an external database

Connecting to databases can be done using ODBC bindings for Lua, or native Lua drivers for databases.

Another way to connect to databases—the preferred way—is to create HTTP queries from Lua and leave the HTTP server (or servers) the burden of managing connection pools, scalability, concurrence, and so on with tried and true well-known techniques.

Here, we'll see this preferred way.

Getting ready

Create a dialplan extension "12348" that will call our last Lua script. Edit the `/usr/local/freeswitch/conf/dialplan/default/04_database.xml` file:

```
<include>
  <extension name="Connect to DataBase">
    <condition field="destination_number" expression="^(12348)$">
        <action application="answer"/>
        <action application="sleep" data="1000"/>
        <action application="lua" data="database.lua"/>
    </condition>
  </extension>
</include>
```

Save the file and then, from `fs_cli`, `reloadxml`.

Also, we need our HTTP server to query the database and send the results to the web clients. As an example, here is a completely unsecure, dangerous, and non-scalable CGI that will query a MySQL database on a Linux machine. Edit the `db.sh` file, make it readable and executable to the HTTP server's user, and put it into the `cgi-bin` directory of the server (on Debian Jessie, it is `/usr/lib/cgi-bin/`):

```
#!/bin/bash
#https://marc.waeckerlin.org/computer/blog/parsing_of_query_string_in_
bash_cgi_scripts
#Decodes an URL-string
function urlDec() {
  local value=${*//+/%20}                      # replace +-spaces by %20
(hex)
    for part in ${value//%/ \\x}; do           # split at % prepend \x
for printf
      printf "%b%s" "${part:0:4}" "${part:4}" # output decoded char
    done
}

echo "Content-type: text/plain"
echo ''

QUERY=$(urlDec "${QUERY_STRING#*=}")
# echo "QUERY = $QUERY"
# echo ''

/usr/bin/mysql -u root --password=rootpwd -B -N -e "$QUERY;"
mysql 2>&1
```

How to do it...

Edit the `/usr/local/freeswitch/scripts/database.lua` file:

```
-- create the api object, to be used to send fs_cli like commands
api = freeswitch.API()

-- pause half a second
session:sleep(500)

-- do an http query to the database and store the result in "how_many_
users" Lua variable
how_many_users=api:execute("curl", "http://192.168.1.125/cgi-bin/db.sh
?query=select+count(*)+from+user")

-- print variable to console and logfile
```

```
session:consoleLog("warning", "how_many_users is: |" .. how_many_users
.. "|\n")

-- trim ending carriage return
how_many_users = string.sub(how_many_users,1,-2)

-- print variable to console and logfile
session:consoleLog("warning", "how_many_users is: |" .. how_many_users
.. "|\n")

-- check number
if(tonumber(how_many_users) < 2) then
        -- play a sound file
        session:execute("playback", "ivr/ivr-there_are.wav")
else
        -- play a sound file
        session:execute("playback", "ivr/ivr-there_is.wav")
end

-- "pronounce" variable to user
session:say(how_many_users, "en", "number", "pronounced")

-- pause half a second
session:sleep(500)

--End of script, automatically hangup
```

How it works...

The working of this recipe can be elaborated as follows:

1. The extension we just created and loaded into the dialplan will match a destination number "12348", answer the incoming call, and then pass the call to the `database.lua` script.

2. We pause for half a second before making an HTTP call to our web server, with the SQL query URL encoded.

3. We put the HTTP server's answer in the `how_many_users` variable.

4. The next line prints the variable's content. This variable will contain the trailing carriage return that ends the string.

   So, we use the `string.sub` Lua function to extract a string that will start from the first character of the original string, and end at two characters from the end of the original string. In simple words, we have just trimmed the trailing carriage return.

5. Then we print it again, and based on its numeric value, we decide which audio file to play to the caller.

6. Next, we pronounce the variable as an English language number, wait for half a second, and exit the script, hanging up automatically.

This is what a call to "**12348**" will look like in the `fs_cli` console (notice that the first time it prints the value of `how_many_users`, it contains a trailing carriage return):

```
freeswitch@internal>
2015-07-04 13:19:28.187830 [NOTICE] switch_channel.c:1089 New Channel sofia/internal
/1002@192.168.1.125 [eff6edd9-6de8-4399-bd31-8f4e7ff45a17]
2015-07-04 13:19:28.227840 [INFO] mod_dialplan_xml.c:642 Processing giovanni <1002>-
>12348 in context default
2015-07-04 13:19:28.227840 [NOTICE] sofia_media.c:92 Pre-Answer sofia/internal/1002@
192.168.1.125!
2015-07-04 13:19:28.227840 [NOTICE] mod_dptools.c:1292 Channel [sofia/internal/1002@
192.168.1.125] has been answered
2015-07-04 13:19:33.787761 [WARNING] switch_cpp.cpp:1256 how_many_users is: |5
|
2015-07-04 13:19:33.787761 [WARNING] switch_cpp.cpp:1256 how_many_users is: |5|
2015-07-04 13:19:35.547773 [NOTICE] switch_core_state_machine.c:315 sofia/internal/1
002@192.168.1.125 has executed the last dialplan instruction, hanging up.
2015-07-04 13:19:35.547773 [NOTICE] switch_core_state_machine.c:317 Hangup sofia/int
ernal/1002@192.168.1.125 [CS_EXECUTE] [NORMAL_CLEARING]
2015-07-04 13:19:35.547773 [NOTICE] switch_core_session.c:1657 Session 41 (sofia/int
ernal/1002@192.168.1.125) Ended
2015-07-04 13:19:35.547773 [NOTICE] switch_core_session.c:1661 Close Channel sofia/i
nternal/1002@192.168.1.125 [CS_DESTROY]
freeswitch@internal>
```

There's more...

Lua can connect to databases in many different ways. Check out FreeSWITCH's Confluence page (`https://freeswitch.org/confluence/display/FREESWITCH/FreeSWITCH+Explained`) for full documentation.

Index

Thank you for buying
FreeSWITCH 1.6 Cookbook

About Packt Publishing

Packt, pronounced 'packed', published its first book, *Mastering phpMyAdmin for Effective MySQL Management*, in April 2004, and subsequently continued to specialize in publishing highly focused books on specific technologies and solutions.

Our books and publications share the experiences of your fellow IT professionals in adapting and customizing today's systems, applications, and frameworks. Our solution-based books give you the knowledge and power to customize the software and technologies you're using to get the job done. Packt books are more specific and less general than the IT books you have seen in the past. Our unique business model allows us to bring you more focused information, giving you more of what you need to know, and less of what you don't.

Packt is a modern yet unique publishing company that focuses on producing quality, cutting-edge books for communities of developers, administrators, and newbies alike. For more information, please visit our website at www.packtpub.com.

About Packt Open Source

In 2010, Packt launched two new brands, Packt Open Source and Packt Enterprise, in order to continue its focus on specialization. This book is part of the Packt open source brand, home to books published on software built around open source licenses, and offering information to anybody from advanced developers to budding web designers. The Open Source brand also runs Packt's open source Royalty Scheme, by which Packt gives a royalty to each open source project about whose software a book is sold.

Writing for Packt

We welcome all inquiries from people who are interested in authoring. Book proposals should be sent to author@packtpub.com. If your book idea is still at an early stage and you would like to discuss it first before writing a formal book proposal, then please contact us; one of our commissioning editors will get in touch with you.

We're not just looking for published authors; if you have strong technical skills but no writing experience, our experienced editors can help you develop a writing career, or simply get some additional reward for your expertise.

FreeSWITCH 1.2

ISBN: 978-1-78216-100-4   Paperback: 428 pages

Build robust, high-performance telephony systems using FreeSWITCH

1. Learn how to install and configure a complete telephony system of your own, even if this is your first time using FreeSWITCH.

2. In-depth discussions of important concepts like the dialplan, user directory, NAT handling, and the powerful FreeSWITCH event socket.

3. Best practices and expert tips from the FreeSWITCH experts, including the creator of FreeSWITCH, Anthony Minessale.

FreeSWITCH Cookbook

ISBN: 978-1-84951-540-5   Paperback: 150 pages

Over 40 recipes to help you get the most out of your FreeSWITCH server

1. Get powerful FreeSWITCH features to work for you.

2. Route calls and handle call detailing records.

3. Written by members of the FreeSWITCH development team.